A Practical Guide to Optical Microscopy

A Practical Guide to Optical Microscopy

John Girkin

Department of Physics, Durham University

CRC Press
Taylor & Francis Group
Boca Raton London New York

CRC Press is an imprint of the
Taylor & Francis Group, an **informa** business

CRC Press
Taylor & Francis Group
6000 Broken Sound Parkway NW, Suite 300
Boca Raton, FL 33487-2742

© 2020 by Taylor & Francis Group, LLC
CRC Press is an imprint of Taylor & Francis Group, an Informa business

No claim to original U.S. Government works

Printed on acid-free paper

International Standard Book Number-13: 978-1-138-06470-6 (Paperback)
International Standard Book Number-13: 978-1-138-06506-2 (Hardback)

Visit the Taylor & Francis Web site at
http://www.taylorandfrancis.com

and the CRC Press Web site at
http://www.crcpress.com

Contents

Preface, xiii

Acknowledgements, xv

About the Author, xvii

CHAPTER 1 ▪ Introduction   1

1.1   A HISTORICAL PERSPECTIVE   2

1.2   INITIAL CONSIDERATIONS ON WHICH METHOD TO CHOOSE
      FOR A SPECIFIC APPLICATION   6

1.3   HOW TO USE THIS BOOK   10

REFERENCES   10

CHAPTER 2 ▪ Understanding Light in Optical Microscopy   13

2.1   INTRODUCTION TO THE PHYSICS OF OPTICAL MICROSCOPY   13

      2.1.1   Light   13

      2.1.2   Waves, Wavelength, Frequency and Particles   15

      2.1.3   Refractive Index   15

      2.1.4   Polarization   16

2.2   THE OPTICS OF MICROSCOPY   17

      2.2.1   Refraction   17

      2.2.2   Interference and Diffraction   19

2.3   LIMITATIONS IN OPTICAL MICROSCOPY   21

      2.3.1   Chromatic Aberration   22

      2.3.2   Spherical Aberration   22

      2.3.3   Astigmatism   24

      2.3.4   Field Curvature   24

2.4   CONTRAST MECHANISMS   24

      2.4.1   Absorption   25

	2.4.2	Scattering	25
	2.4.3	Phase or Refractive Index Changes	27
	2.4.4	Fluorescence	27
	2.4.5	Fluorescence Lifetime	28
	2.4.6	Non-linear Excitation; Harmonic Generation, Raman Scattering	28
2.5	A BRIEF INTRODUCTION TO LIGHT SOURCES FOR MICROSCOPY		28
	2.5.1	Conventional Filament Bulb (Including Metal Halide)	29
	2.5.2	Mercury Vapour Bulb	29
	2.5.3	Light Emitting Diodes (LEDs)	29
	2.5.4	Arc Lamp	29
	2.5.5	Lasers	30
2.6	DETECTION OF LIGHT IN MICROSCOPY		30
	2.6.1	Human Eye and Photographic Film	30
	2.6.2	Electronic or Digital Camera	32
	2.6.3	Photomultiplier	34
	2.6.4	Photodiodes	36
REFERENCES			36

CHAPTER 3 ■ Basic Microscope Optics — 39

3.1	INTRODUCTION	39
3.2	BASIC TYPES OF WIDEFIELD OPTICAL MICROSCOPE	39
3.3	CORE OPTICS OF A WIDEFIELD MICROSCOPE	42
3.4	OPTIMAL ILLUMINATION	44
3.5	MICROSCOPE OBJECTIVES	46
3.6	IMAGE DETECTION AND RECORDING	50
3.7	GUIDELINES FOR USE AND ADVANTAGES AND DISADVANTAGES OF A WIDEFIELD MICROSCOPE	51

CHAPTER 4 ■ Advanced Widefield Microscopy — 55

4.1	INTRODUCTION		55
4.2	POLARIZATION MICROSCOPY		55
	4.2.1	Practical Implementation of Polarization Microscopy	56
	4.2.2	Applications of Polarization Microscopy	58
4.3	PHASE CONTRAST MICROSCOPY		58
	4.3.1	Practical Implementation of Phase Contrast Microscopy	59
	4.3.2	Applications of Phase Microscopy	61

4.4 DIFFERENTIAL INTERFERENCE CONTRAST (DIC) MICROSCOPY 62

4.4.1 Practical Implementation of DIC Microscopy 64

4.4.2 Applications of DIC Microscopy 65

4.5 DARKFIELD MICROSCOPY 65

4.5.1 Practical Implementation of Darkfield Microscopy 66

4.5.2 Practical Applications of Darkfield Microscopy 68

4.6 FLUORESCENCE MICROSCOPY 68

4.6.1 Practical Implementation of Fluorescence Microscopy 68

4.6.2 Practical Applications of Fluorescence Microscopy 72

4.7 SUMMARY AND METHOD SELECTION 72

CHAPTER 5 ■ Confocal Microscopy 73

5.1 PRINCIPLES OF CONFOCAL MICROSCOPY 74

5.2 BEAM SCANNED CONFOCAL SYSTEM 75

5.3 FILTER SELECTION FOR BEAM SCANNED CONFOCAL SYSTEMS 79

5.4 DETECTOR SELECTION FOR BEAM SCANNED CONFOCAL SYSTEMS 80

5.5 NIPKOW OR SPINNING DISK CONFOCAL SYSTEMS 82

5.6 PRACTICAL GUIDELINES TO MAXIMIZE THE PERFORMANCE OF A CONFOCAL MICROSCOPE 83

5.6.1 Microscope Choice, Sample Mounting and Preparation 85

5.6.2 Lens Selection 87

5.6.3 Initial Image Capture 88

5.6.4 Optimization 90

5.6.5 Saving Data 93

5.6.6 Routine Maintenance 94

5.7 RECONSTRUCTION 95

REFERENCES 95

CHAPTER 6 ■ Fluorescence Lifetime Imaging Microscopy (FLIM) 97

6.1 INTRODUCTION TO FLUORESCENCE LIFETIME 97

6.1.1 Absorption and Emission 97

6.1.2 Fluorescence Lifetime 100

6.2 MEASUREMENT TECHNIQUES 101

6.2.1 Time Correlated Single Photon Counting (TCSPC) 101

6.2.2 Time Gating Electronics 103

6.2.3 Time Gated Camera 104

6.2.4 Phase Measurement 106

6.2.5 Slow Detector Method 107

6.3 METHODS OF ANALYSIS 109

6.4 EXPERIMENTAL CONSIDERATIONS AND GUIDELINES FOR USE 111

6.4.1 FLIM for Enhancing Contrast 112

6.4.2 FLIM for FRET and Observing Changes in Lifetime 112

6.4.3 FLIM for Absolute Lifetime Measurement 114

6.4.4 Practical Considerations 115

REFERENCES 116

CHAPTER 7 ■ Light Sheet or Selective Plane Microscopy 117

7.1 INTRODUCTION 117

7.2 BRIEF HISTORY OF LIGHT SHEET MICROSCOPY 119

7.3 OPTICAL PRINCIPLES 120

7.4 PRACTICAL SYSTEMS 123

7.4.1 Optical Details 123

7.4.2 Basic Alignment 125

7.4.3 Basic Variations in SPIM 126

7.5 PRACTICAL OPERATION 127

7.5.1 Sample Mounting 127

7.5.2 Basic Operation 129

7.5.3 Correcting Common Faults 130

7.5.4 High Speed Imaging and Synchronization 131

7.5.5 High Throughput SPIM 132

7.6 SPIM IMAGING PROCESSING 133

7.7 ADVANCED SPIM METHODS 134

7.8 WHAT IS SPIM GOOD FOR AND LIMITATIONS 135

REFERENCES 136

CHAPTER 8 ■ Multiphoton Fluorescence Microscopy 139

8.1 INTRODUCTION TO MULTIPHOTON EXCITATION 140

8.1.1 Requirements on Light Sources for Multiphoton Excitation 142

8.1.2 Requirements on Photon Detection for Multiphoton Microscopy 143

8.2 PRACTICAL MULTIPHOTON MICROSCOPY 145

8.2.1 Wavelength 145

8.2.2 Pulse Width and Dispersion Compensation 146

8.2.3	Average Power	151
8.2.4	Objective Lens Selection	152
8.2.5	Detection	153
8.3	GOING FROM CONFOCAL TO MULTIPHOTON MICROSCOPY	154
8.4	ADVANCED MULTIPHOTON MICROSCOPY	157
8.4.1	Endoscopic Multiphoton Microscopy	157
8.4.2	Adaptive Optics for Aberration Correction	159
8.4.3	Measurement of Two Photon "Dose"	160
REFERENCES		162

CHAPTER 9 ■ Harmonic Microscopy		165
9.1	PHYSICAL BASIS FOR HARMONIC GENERATION	166
9.2	PRACTICAL HARMONIC MICROSCOPY	168
9.3	APPLICATIONS OF HARMONIC MICROSCOPY	171
REFERENCES		173

CHAPTER 10 ■ Raman Microscopy		175
10.1	PHYSICAL BASIS OF THE RAMAN EFFECT	176
10.2	COHERENT ANTI-STOKES RAMAN SCATTERING (CARS)	179
10.3	STIMULATED RAMAN SCATTERING (SRS) MICROSCOPY	181
10.4	PRACTICAL RAMAN MICROSCOPY INSTRUMENTATION	181
10.5	PRACTICAL CARS MICROSCOPY INSTRUMENTATION	185
10.6	TECHNIQUES AND APPLICATIONS IN RAMAN MICROSCOPY	186
10.7	TECHNIQUES AND APPLICATIONS IN CARS AND SRS MICROSCOPY	189
10.8	WHEN TO CONSIDER THE USE OF RAMAN MICROSCOPY	191
REFERENCES		191

CHAPTER 11 ■ Digital Holographic Microscopy		193
11.1	PHYSICAL BASIS OF THE METHOD	194
11.2	PRACTICAL IMPLEMENTATION	197
11.2.1	The Light Source for Holographic Microscopy	198
11.2.2	The Detector for Holographic Microscopy	199
11.2.3	The Reconstruction Algorithm for Holographic Microscopy	200
11.3	PRACTICAL APPLICATIONS OF DIGITAL HOLOGRAPHIC MICROSCOPY	201
11.3.1	Surface Microscopy	202

11.3.2 Particle Tracking 202

11.3.3 Cell Imaging 203

11.3.4 Total Internal Reflection Digital Holographic Microscopy 203

11.4 SUMMARY 205

REFERENCES 205

CHAPTER 12 ▪ Super Resolution Microscopy 207

12.1 INTRODUCTION 207

12.2 TOTAL INTERNAL REFLECTION MICROSCOPY 208

12.2.1 Principles of Total Internal Reflection Microscopy 208

12.2.2 Practical Implementations of Total Internal Reflection Microscopy 211

12.2.3 Practical Considerations for Total Internal Reflection Microscopy 211

12.3 STRUCTURED ILLUMINATION MICROSCOPY (SIM) 212

12.3.1 Principles of Structured Illumination Microscopy 213

12.3.2 Practical Implementations of Structured Illumination Microscopy 214

12.3.3 Practical Considerations for Structured Illumination Microscopy 215

12.4 LOCALIZATION MICROSCOPY (STORM/PALM) 217

12.4.1 Principles of Localization Microscopy 217

12.4.2 Practical Implementations of Localization Microscopy 219

12.4.3 Practical Considerations for Localization Microscopy 219

12.5 STIMULATED EMISSION AND DEPLETION (STED) MICROSCOPY 220

12.5.1 Principles of STED 220

12.5.2 Practical Implementations of STED 221

12.5.3 Practical Considerations for STED 222

12.6 SELECTION OF SUPER-RESOLUTION METHODS 222

REFERENCES 223

CHAPTER 13 ▪ How to Obtain the Most from Your Data 225

13.1 INTRODUCTION 225

13.2 BASICS OF DATA COLLECTION 226

13.3 SOFTWARE CONSIDERATIONS 227

13.3.1 Open Source Image Processing Packages 228

13.3.2 Programming Languages for Image Processing 228

13.3.3 Commercial Image Processing Packages 229

13.4 BASICS OF DATA PROCESSING 231

13.4.1 Core Techniques 231

13.4.1.1 *Reducing Noise* 231

13.4.1.2 *Uneven Illumination* 232

13.4.1.3 *Increasing Contrast* 233

13.4.1.4 *Enhancing Perceived Detail* 235

13.4.1.5 *Monochrome Look-Up Tables and the Addition of Colour* 236

13.5 PRODUCING QUANTIFIED DATA FROM IMAGES 236

13.5.1 Intensity-Based Quantification 238

13.5.2 Spatial-Based Quantification 238

13.5.3 Temporal-Based Quantification 239

13.6 DECONVOLUTION 240

13.7 SUMMARY 242

REFERENCES 242

CHAPTER 14 ■ Selection Criteria for Optical Microscopy 243

14.1 INTRODUCTION 243

14.2 BASIC SELECTION GUIDELINES 244

14.3 SPECIALIZED TECHNIQUES 248

14.3.1 Fluorescence Recovery after Photobleaching (FRAP) 248

14.3.2 Förster Resonant Energy Transfer (FRET) 251

14.3.3 Opto-genetics, Observation of Cell Ablation and Photo-uncaging 252

14.3.4 Imaging of Plants and Plant Cells 253

14.4 SUMMARY 253

REFERENCES 254

GLOSSARY, 255

INDEX, 259

Preface

I WAS GIVEN MY FIRST, very simple, microscope at the age of around eight and although I was interested in observing the microscopic world it was how this was achieved that interested me even more. At a slightly later age I also remember disassembling my father's old binoculars to look at the way that they operated. Using sunlight through a curtain to produce a beam of light, similar to some of the diagrams in Newton's various notebooks that I had seen reproduced, I was able to observe the colours and effects of the different optical elements. This desire to understand how things work, and how they can be improved practically, has been the main focus of my career. When I was introduced to the laser at university, and the remarkable versatility of this light source, I was then hooked on using light both as a tool itself in high resolution laser spectroscopy (the subject of my PhD) and also as a method to develop a wide range of instrumentation. This work has varied from optically based non-contact tonometers (for measuring the pressure in the eye), through to optical tweezers to mimic a dishwasher under a microscope. At the core of all this work has been the desire to make practical, optically based instruments to enable others to really understand nature or to improve clinical practice.

The genesis of this book has come through multiple collaborations with users of optical microscopy, frequently life scientists, who really want to push optical methods for the insight and understanding they bring to their specific scientific questions. As a result of these interactions I was frequently asked to deliver presentations on both basic and advanced optical microscopy, with the audience ranging from undergraduates through to senior and internationally prominent academics. The audience expertise covered life scientists wishing to obtain a better understanding of how microscopes worked, as well as physicists interested in applying their optical expertise to build better instruments. It also became increasingly clear that although there were a number of books dedicated to optical microscopy, these frequently focused on specific methods, or were highly specialized in their application.

This book is aimed at being a practical guide, with a role not only in helping to develop research protocols, but also as a book that is placed by the microscope in the laboratory so that the best possible use is made of the instrument. The optical physics within the text is kept to a minimum, but there are inserts, which can be avoided, which provide the physical basis for the effects described. The core aim is to encourage the use of the correct microscope for each specific imaging task. Wonderful images can now be captured using very low cost "web cam" based microscopes and these should not be overlooked as the best instrument to be used just because there is also a non-linear microscope available. The text is written in a style that provides the basic optical physics that needs to be known to understand the specific imaging modality. It is frequently said that physics is taught by diminishing deception, that as one moves through the academic system one appreciates what one was taught at school was not the entire truth. In this spirit I have attempted to use as basic an explanation as possible in the main body of the text with the full technical detail in the inserts. I hope this makes the book approachable by the non-expert while still providing the detailed information required by others. Each chapter can be read alone though it is suggested that those with little physics or optical background might read Chapter 2 before looking at any specific technique chapter, as this will provide them with an understanding of the way that all optical systems operate, and how light interacts with matter to produce microscopic images.

One challenge in writing this book is that the field is constantly changing with more advanced and specific methods being developed all the time. However, the fundamentals of all optical microscopy methods are covered and thus if a new, highly specialized method does emerge, I hope that by reading the relevant pages in this book and then the relevant academic publication the reader will be able to appreciate the latest innovation. Throughout I have tried to remember that the optical microscope is the tool to help develop greater understanding of a specific challenge and that collecting the most suitable data should always be the focus of any imaging method.

Acknowledgements

I WOULD FIRST LIKE TO thank my parents for their unfailing encouragement and support while I was growing up: without this I would never have even started my career, let alone this book. I would like to thank my wife Joan and sons Michael and Nicholas for their support during the writing process, for their encouragement and help to improve the quality of the text. I must also acknowledge the huge efforts provided by my editor Rebecca Davies and editorial assistant Kirsten Barr. Their dogged determination and discipline, reminding me when chapters were due, really saw this book through to completion. Any mistakes present are entirely down to me, not this wonderfully supportive team.

About the Author

John **Girkin** is Professor of Biophysics at Durham University and Director of the Biophysical Sciences Institute in Durham, UK. Originally trained as a physicist at Oxford and with a PhD from Southampton University in Laser Spectroscopy of Atomic Hydrogen he worked for ten years in industry before returning to academia at the Institute of Photonics, Strathclyde University. He has published over 100 peer-reviewed publications, is a Fellow of both the Institute of Physics and the Optical Society of America serving on both national and international review panels in the area of biophotonics and microscopy.

CHAPTER 1

Introduction

O PTICAL MICROSCOPES ARE PROBABLY the most widely used instrumentation across all branches of science and medicine, and play a significant role in helping to advance human understanding through research as well as being an everyday tool for quality control in industry and diagnosis in the clinic. They perhaps provide the ultimate implementation of the idiom "seeing is believing" as microscopy has performed a crucial part in helping us explore and understand a world where structures and organisms are smaller than can be seen directly with the human eye. For hundreds of years optical microscopes and telescopes have paved the way for new scientific insights at both ends of the length scale. The desire for ever higher quality images and improved resolution has led to significant advances in both theoretical optical physics and practical engineering and, as new photonics technology has been invented, this has rapidly been applied in optical microscopy. In the last twenty years there has been a significant upsurge in developments in the field of optical microscopy leading to a vast array of new imaging modalities, each of which is excellent for one imaging task but may be less suitable for another. The aim of this book is to provide a simple guide in helping to select the most suitable method of optical microscopy to use for a particular application and the subsequent use of the preferred instrumentation.

To make the book as useful as possible to a wide range of readers each chapter is self-contained with the physical concepts behind a specific imaging process being explained through marked areas of text. In order to appreciate how a method is best applied to a particular task it is not necessary to read these more detailed physical insights. Although the text is generally focused on the application of microscopy to biological samples the general principles described are broadly applicable and where possible suitable examples from geology and materials science are given. Although huge technological advances in optical microscopy methods have been made in the last twenty years the core "physics of imaging" has not changed and the limitations that these play in microscopy will be highlighted throughout. An appreciation of these principles is very helpful to understand the trade-offs that always have to be made between spatial resolution, speed of imaging and the level of perturbation to the sample in order to obtain the images required. The underlying ethos used throughout is that of selecting the correct tool for the particular task; if results can be achieved with a low-cost camera and magnifying lens then this is the route to select. The highly complex and technical methods can then be applied to the tasks that really require that level of sophistication. However, before presenting the details of both the imaging process and subsequent practical implementation of the methods it is worth considering optical microscopy in a historical perspective.

1.1 A HISTORICAL PERSPECTIVE

Being able to observe events has always been the cornerstone of scientific discovery ranging from monitoring the movement of the sun, moon and stars through to watching the tiniest visible creatures navigate their way around the world. This desire to see events at different scales has, throughout the centuries, advanced optical instrumentation and, as will be shown shortly, provided a significant boost to basic understanding of light and optics. Such visual observations were initially undertaken using just the human eye. Since the invention of the lens around 2000 years ago, there has been a slow refinement in the ability to image with ever greater detail at both astronomical and miniature scales. Perhaps the first recorded use of the lens for making observations at a microscopic scale was by Pliny the Elder (in Bostock, 1855) where a magnifying lens was used to look at tissue damage at a fine scale and also, in some cases, to cauterize the wounds using sunlight. At this stage clearly the optical physics taking place to produce the magnified image was not understood and the lenses were produced through the controlled melting of glass to form droplets, with luck as well as skill being involved in the final quality of the product. Although observations on the effect of beads of glass, and indeed water filled containers, were made by scholars such as Ibn al-Haytham and Roger Bacon, practical applications of the magnifying effect were not really considered, or at least recorded, until Salvino D'Armate in Italy during the 12th century. Using a single positive lens, which he was making to help correct the loss of close sight with age (presbyopia), he noted that the world could be enlarged to the viewer. There is however little further information on what was observed in this new "microscopic" world.

Around 300 years later, in 1590, when the craft of spectacle making was firmly established, Zacharias Jansen and his father Hans, in the Netherlands, created what is generally viewed as being the first optical microscope. This was a compound instrument consisting of multiple lenses developed entirely through experimentation and carefully recorded observations. Although all their original notes were lost during the Second World War it is known they developed an instrument that consisted of three tubes which were moved for focusing and altering the magnification (Figure 1.1A). The eyepiece was a bi-convex lens and the objective a plano-convex lens, which required the most advanced polishing methods available in the late 16th century. Perhaps this is the first example of the microscope making use of an optical technological advance. Based upon the notes made the magnification could be adjusted to be between three and ten times, depending on the position of the sliding tubes and working distance. Although crude by modern standards, both in terms of optical design and fabrication, it opened up the world of the very small for the first time. It is known that Galileo Galilei also was experimenting with compound optical systems to develop his first telescope and was aware of the advances being made in the Netherlands. In 1609, he developed a compound microscope using both convex and concave lenses.

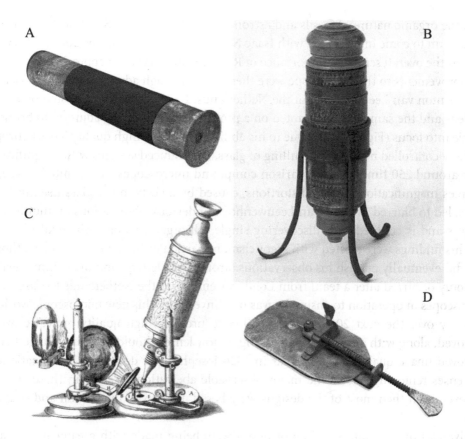

FIGURE 1.1   A) One of the original Jansen microscopes, B) Galileo Galilei's original microscope, C) Drawing of Hooke's microscope; note the oil lamp and large lens for improved illumination, D) A copy of Anton van Leeuwenhoek's single lens, ×200 magnification microscope. (Images A and D credit to Molecular Expressions.com at Florida State University, Image B Credit Museo Galileo).

He described his "modified telescope to see objects very close" in his book *The Assayer* (*Il Saggiatore*) (1623) and several variations of this instrument were built (Figure 1.1B). It was around this time that the term "microscope" was first used.

Now that the basic compound microscope had been developed, for the next few hundred years it was the application of this new instrument that became increasingly important as people started to look at a wide range of samples. Robert Hooke, while the Curator of Experiments of the Royal Society, published his findings (1665) in which the term "cell" was used for the first time to describe what he thought of as the basic unit of life. The images contained in his book, *Micrographia*, were taken with one of the best microscopes of the time and the illustration of its operation (Figure 1.1C) demonstrates the importance of the illumination system occupying almost as much of the image as the actual observation optics. It is perhaps interesting to note here that for many of the most advanced systems in use today described in the following chapters the light source still occupies a major physical area in the laboratory! Although best known for the high quality copperplate engravings (from Hooke's original drawings) the book also describes the wave theory of light, the organic nature of fossils and astronomical observations. Some of these writings caused him to come into conflict with Isaac Newton and it is only in the 20th century that perhaps the overall scientific importance of Robert Hooke has been appreciated.

Improvements to the microscope were then made through advances in lens fabrication led by Anton van Leeuwenhoek in the Netherlands. His microscope consisted of a single ball lens and the sample was mounted on a pin that could be finely adjusted to bring the sample into focus (Figure 1.1D). Due to his ability to produce high quality lenses, through carefully controlled melting and pulling of glass, he produced systems with magnification up to around 250 times. By comparison compound microscopes were limited to around 30 times magnifications before distortions, caused by defects in the glass used in larger lenses, led to blurred images. Van Leeuwenhoek undertook observations on multiple live samples and is credited with discovering single cell organisms or "animalcules". At the time his findings were treated with scepticism, in particular by the Royal Society, though they did eventually publish his observations (around 560 letters) and made him a Fellow. This only occurred after a team from London went out to the Netherlands to observe his microscopes in operation to ensure he was not "inventing" this new microscopic world!

Slowly over the next 200 years the ability to produce high quality optical elements improved, along with a basic understanding of how lenses should be brought together for improved image quality. In particular in 1826 Joseph Lister developed achromatic doublet lenses removing one of the major observable aberrations present in these systems. However, even then most of the designs were based upon experimentation and trial and error.

Advances in the understanding of optics were being made with greater understanding of the mathematics behind light propagation and the behavior of waves. However, the whole field of optics was then revolutionized through trying to build better microscopes. Ernst Abbe working for Carl Zeiss in Jena developed the framework for the full performance, and limitation, of optical imaging instruments (Abbe 1873) and later presented his findings to the Royal Microscopical Society (Abbe 1881). Abbe's findings were largely based upon careful experimentation while around the same time Herman Helmholtz was working on a theoretical approach. Joseph Lagrange, more than 60 years previously, had hinted at the limitations through mathematically based reasoning. As a result of this discovery Abbe became a shareholder in the Zeiss company becoming very wealthy, though he did also introduce the eight-hour day into the Zeiss workshop. He proved that the maximum resolution possible is proportional to the wavelength and inversely proportional to the numerical aperture (light gathering power) of the lens. Details on the physical aspects of this fundamental limitation are provided in Chapter 2. Abbe's work at this time advanced the entire optical field for practical devices and instruments and although huge technical innovations have been made in optical microscopes since this date the majority of improvements have been through technology and engineering rather than fundamental changes in optical design.

In 1902 Richard Zsigmondy was interested in the study of colloidal gold suspensions and wanted to observe the movement and position of particles that were smaller than Abbe's limit of resolution. Working with Henry Siedentopf (an employee of Zeiss) he developed a microscope in which a sheet of light illuminated the sample and the observation of the light scattered by the particles was observed orthogonally to the excitation. This technique of "ultramicroscopy" (Siedentopf and Zsigmondy 1902) was a method really ahead of its time as the observations all had to be made by eye since photography was not sufficiently advanced and CCD cameras not even a dream. However, the results of Zsigmondy's observations were sufficient for him to be awarded the Nobel Prize for Chemistry in 1925. The method was rediscovered as light sheet or selective plane illumination microscopy (Huisken et al. 2004). This preferred method of imaging *in vivo* dynamic events especially for extended periods of time is described in Chapter 7.

Subsequent developments in basic optical microscopy have generally been in methods of improving the contrast in the image rather than significant changes in optical design. In this context the next major advance was in the development of phase contrast microscopy by Fritz von Zernike in 1934 (Zernike 1934) which is described in detail in Chapter 4. Zernike was awarded the 1953 Nobel Prize for Physics for this breakthrough and he acknowledged the role that Abbe's work had played in his discovery. As other methods of seeing minute structures developed, such as electron microscopy, with resolution that was better than that offered by optical methods, people started to consider what the real advantages were in optical microscopy. This should still be the first question asked when considering how to observe a very small structure or process.

A major benefit of the optical microscope is clearly the ease of use, at least for basic systems, alongside the simple sample preparation without the need for complex fixing and cutting of ultra-thin sections, and this provides the ability to image live samples and dynamic processes. This shift towards live and dynamic samples, although Leeuwenhoek did image live sperm on the pin of his microscope, created an interest in developing optical microscopes with the ability to image in three dimensions. In 1961 Marvin Minsky patented a design of a confocal microscope capable of taking optical sections through intact tissue (Minsky 1961), though this is a clear example of an idea before its time. Lasers as a source of bright light had only just been developed and were still very much research tools in the physicist's laboratory. Beam scanning of optics and the electronic recording of data were also both very much in their infancy. In Minsky's microscope, developed to look at neurological samples, the light was provided by an arc lamp and the detected signals were displayed on a long persistence oscilloscope linked to the slow mechanical scan of a mirror. A camera was attached to the oscilloscope and a long exposure taken as the entire optical section was scanned to produce one photograph. Not a very practical system and it was only in the 1970s that confocal systems were developed in the laboratory based upon sample scanning. Then in the 1980s practical confocal microscopes first became commercially available through BioRad, based on a beam scanning design and developed in Cambridge by John White and Brad Amos (White and Amos 1987). Crucial to this practical optical development was the growing availability and performance of low cost personal computers and this, linked with the advent of digital imaging sensors and novel light sources, has led to a revolution in optical microscopy. Two-photon microscopy was then reported (Denk, Strickler and Webb 1990), where the fluorescent excitation was limited in three dimensions through the use of non-linear absorption of multiple photons.

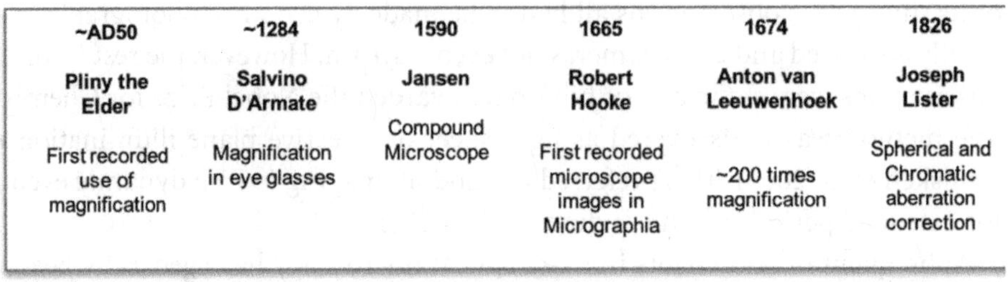

~AD50	~1284	1590	1665	1674	1826
Pliny the Elder	Salvino D'Armate	Jansen	Robert Hooke	Anton van Leeuwenhoek	Joseph Lister
First recorded use of magnification	Magnification in eye glasses	Compound Microscope	First recorded microscope images in Micrographia	~200 times magnification	Spherical and Chromatic aberration correction

FIGURE 1.2   A timeline of major optical microscope developments.

Numerous variations on this method were then reported in which other multi-photon processes were used to localize excitation in the sample including second and third harmonic imaging and Coherent Raman excitation. The recent culmination of these advances might be viewed as being the award of the 2014 Nobel Prize for Chemistry for methods of beating Abbe's diffraction limit. Here combinations of localized excitation, novel photochemistry, advanced optical sources, new detectors and high performance computing are all brought together for improved resolution microscopy. All of these recent advances, as well as the better-established methods, are subsequently described in the following chapters. Figure 1.2 illustrates some of the key milestones in the development of optical microscopy and it is interesting to note that the horizontal "time scale" is not linear, illustrating the numerous recent developments.

1.2 INITIAL CONSIDERATIONS ON WHICH METHOD TO CHOOSE FOR A SPECIFIC APPLICATION

Before examining in detail the different forms of optical microscopy it is worthwhile considering how the vastly different fields in which they are used affects the type of imaging method that might be initially considered. For example, looking at a section of rock or metal is likely to require a different method to that needed to observe changes in the cytoskeleton within a live cell or for looking at processes deep within a living, intact organism. However, there are a few general guidelines that should always be considered, with the overriding principle being to select the simplest method that will provide the required data with the minimal perturbation to the sample. The initial questions to consider are:

1. What do I really want to know about my sample?

2. Is my sample generally transparent and at what wavelengths?

3. Is my sample alive or dead?

1873	1902	1934	1961	1990	1994-
Ernst Abbe	**Richard Zsigmondy**	**Fritz Zernike**	**Marvin Minsky**	**Watt Webb and others**	**Eric Betzig Stefan Hell William Moerner**
Determines factors limiting resolution	Ultra-microscope	Phase Contrast	Determines factors limiting resolution	Non-linear excitation	Super-resolution Techniques

4. What temporal resolution is required?

5. What spatial resolution is required?

6. What is the contrast mechanism required to view the details of interest?

7. Two dimensions or three dimensions?

8. How long do I wish to observe the sample for?

9. How will the sample be mounted?

10. Does the sample require any special handling or local conditions?

11. How large, or small, is the sample?

12. Can the sample tolerate exposure to light?

13. What equipment is readily available?

14. How is the data recorded going to be handled (including storage, analysis and presentation)?

Clearly many of these questions are linked together but they provide a quick guide to what should be considered when selecting an optical microscopy method. Figure 1.3 illustrates some of the considerations that should go into this selection process and how the answers will guide the user towards a specific technique. Later chapters provide the details on how the answers to the questions determine the most suitable method to be adopted for a particular application. For instance, in the field of geology and metallurgy the sample is not likely to be highly transparent, unless a very thin section has been prepared, and thus an epi-imaging system is probably required. In such a system the light is delivered to the sample and the return light collected through the same objective lens. Epi-imaging is also

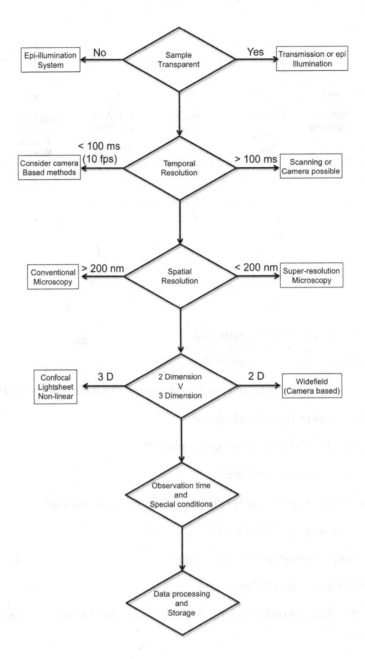

FIGURE 1.3 Outline decision tree to help in selecting the most suitable optical microscopy technique.

possible with transparent samples but in their case it is also possible to illuminate by transmission through the sample.

Keeping a living specimen alive whilst imaging presents practical issues, and minimal perturbation to the sample is a high priority. However, crucially important to all microscopy methods is the contrast mechanism as without contrast nothing can be seen. In general a method of microscopy is described by the contrast mechanism by which the sample is being viewed such as fluorescence, polarization, Raman and second harmonic. All of the current methods of developing contrast to determine features within a sample are explained and discussed in the following chapters alongside ways of maximizing the contrast to produce the highest quality datasets.

The largest single consideration in selecting a particular imaging method is likely to be made of the compromise between spatial resolution and time of imaging. Many of the super-resolution methods (Chapter 12), which beat the Abbe imaging limit, use extended viewing times to improve the resolution. This is through capturing multiple images and post-processing or using temporal and light activation tricks to limit the volume from which the light is collected. Thus, if one wishes to follow dynamic processes within a cell, for example imaging calcium transients, then using a method which requires several seconds to record a single image is unlikely to be suitable; however, on a fixed sample of tissue or geological slice, time is less likely to be an issue.

The next fundamental question is core to many decisions: that of having two- or three-dimensional datasets and in some cases adding extended time viewing. The focus for many recent advances in many application areas, but especially for biological applications, has been to add the ability to image dynamically. This is linked with the development of optical sectioning methods for intact samples, such as confocal or light sheet microscopy, and with advances in labelling intact samples, frequently now through genetic manipulation so that a biological sample can produce its own fluorescent markers. At a deeper level this also integrates with improvements in high performance computing where dynamic models can be developed to help interpret the data produced by four-dimensional experimental datasets. Ideally one would like to observe all processes taking place inside as intact a sample as possible, and clearly in the case of the life sciences this would be in an intact, living organism. While this may be possible in single cell creatures (within the limits imposed for diffraction, etc.) for more complex organisms this clearly becomes much harder. Prior to the invention of optical sectioning microscopes, studying dynamic processes within biology was undertaken by imaging a range of fixed, physically sectioned samples taken at different points, from different samples, and trying to develop a model for the processes taking place. An analogy is that of trying to work out the rules of a sport, say football (soccer), by going to one match and digging a hole in the pitch after one minute (one is perturbing the system radically), pointing a camera in a specific direction and taking a single image. One then goes to a second match, digs another hole, in what you believe is the same place on the pitch, and after two minutes recording a second image, and so on for a full ninety separate minutes at ninety different matches. One then takes all the single images and attempts to determine the rules. Clearly a much better route would be to watch a single match, from above the pitch, and monitor what is taking place. There is a risk that the ball, or perhaps players, might hit the camera but that is clearly less invasive than watching a number of matches in which you have dug holes at fixed time points. Optical microscopy is therefore now helping to reduce the use of multiple animals in studies – an important ethical consideration in all experiments. The advent of optical sectioning microscopy, in all its forms, has now made the ability to image in four dimensions possible, though sometimes with the risk of high light dosage to the sample.

The other questions on the selection of technique mainly relate to minimizing the perturbation to the sample either through its local environment or reducing the light dose. These are likely to be more important for life science applications than for imaging robust metal or rock samples, though even in the latter case the light could affect the local chemistry within a sample.

The final question is perhaps in many ways the most important. Microscopy, in general, is not just used to obtain an attractive image though this may well sometimes be a beneficial by-product. The reason for capturing an image, or more frequently an image series, is to help understand the structure and processes taking place within a sample, and normally this means some form of computer processing of the data. As well as considering what data to record it is also important to consider how such data might be stored and subsequently handled. With some of the extended imaging techniques now available, for example selective plane illumination microscopy (SPIM), full three-dimensional data sets can be recorded every few minutes for perhaps forty-eight hours, or more. This can lead to a single dataset being over 2Tb, and thus consideration of storage, or even just transferring the data from the imaging laboratory to a processing computer needs to be considered. Although advances in optical techniques are still coming at a very rapid rate, the next real technological breakthroughs are likely to be needed in dealing with such datasets. As technology becomes more sophisticated, and multiple imaging modalities are combined, this challenge is only going to grow. These practical issues and further help in guiding the selection of the most suitable method are expanded upon in Chapters 13 and 14.

1.3 HOW TO USE THIS BOOK

The structure of this book is designed to make it easy to use as a practical laboratory guide to optical microscopy. Throughout, the main text of each chapter can be read without the need for specialized optics or physics knowledge. Small extra sections are present which do provide the physical principles behind the text but these are not vital to obtain a good understanding of each method. Each chapter, after Chapter 2, can thus be read and understood in isolation. In Chapter 2 the basic properties of light and imaging systems are considered and although not essential for the following chapters some of the specific terms and physics behind optical microscopy are explained, and may prove to be useful to those without some advanced school level physical science background. Chapter 2 also covers some of the technology that will be used in later chapters such as ultra-short pulse lasers and the different forms of digital imaging sensors now available. Frequently references are made to recent publications in which the imaging methods have been used to provide a more detailed context in which the different techniques can be used. The majority of the applications and examples given are based upon the requirements of life science users, although where relevant examples are given from other fields.

REFERENCES

Abbe, E. 1873. "Beiträge zur Theorie des Mikroskops und der Mikroskopischen Wahrnehmung". *Archives Microscope Anat* 9: 413–18.

Abbe, E. 1881. "On the Estimation of Aperture in the Microscope". *Journal of the Royal Microscopical Society* 1(3): 388–423. http://doi.wiley.com/10.1111/j.1365-2818.1881.tb05909.x.

Denk, W., J. H. Strickler and W. W. Webb. 1990. "Two-Photon Laser Scanning Fluorescence Microscopy". *Science (New York)* 248(4951): 73–6. www.ncbi.nlm.nih.gov/pubmed/2321027.

Galilei, G. 1623. *The Assayer*. Rome.

Hooke, R. 1665. *Micrographia: Or, Some Physiological Descriptions of Minute Bodies Made by Magnifying Glasses*. eds J. Martyn and J. Allestry. London.

Huisken, J., J. Swoger, F. Del Bene, J. Wittbrodt and E. H. Stelzer. 2004. "Optical Sectioning Deep inside Live Embryos by Selective Plane Illumination Microscopy". *Science (New York, N.Y.)* 305(5686): 1007–9. www.ncbi.nlm.nih.gov/pubmed/15310904.

Minsky, M. 1961. "Microscopy Apparatus". www.google.com/patents/US3013467.

Pliny the Elder. 1855. "The Natural History". In *The Natural History, Book XXXVII*, ed John Bostock. London.

Siedentopf, H. and R. Zsigmondy. 1902. "Uber Sichtbarmachung und Größenbestimmung Ultramikoskopischer Teilchen, mit Besonderer Anwendung auf Goldrubingläser". *Annalen der Physik* 315(1): 1–39. http://doi.wiley.com/10.1002/andp.19023150102.

White, J. G. and W. B. Amos. 1987. "Confocal Microscopy Comes of Age". *Nature* 328(6126): 183–4. http://dx.doi.org/10.1038/328183a0.

von Zernike, F. 1934. "Beugungstheorie des Schneidenver-Fahrens und Seiner Verbesserten Form, der Phasenkontrastmethode". *Physica* 1(7): 689–704. http://dx.doi.org/10.1016/S0031-8914(34)80259-5.

CHAPTER 2

Understanding Light in Optical Microscopy

A T THE CORE OF all optical imaging methods is light, and the image seen through a microscope is a result of the way that light interacts with the sample. This chapter outlines the physical principles that underpin optical microscopy without the requirement for a high level of physics expertise. Initially the basic concepts of light are presented; this is followed by an explanation of how optical components manipulate light to provide high quality microscopy images. Optics are not always perfect, and some of the complications in manipulating light and how this can lead to images that are poor in quality, crucially lacking the detail required to understand the processes being studied, are also discussed. However, even the best possible optics will provide no suitable images without contrast, or the ability to highlight the areas of interest from the background. A range of standard contrast mechanisms is therefore explained, illustrating how the different properties of light presented earlier are used to enhance the features of interest in an image. The chapter then finishes by looking at the two ends of an optical microscopy system, the light source and the detector. Throughout the chapter as new terms are introduced these are explained with the minimal use of mathematical formulation. Where applicable the mathematical explanation for a specific process is provided in a text box, but this detail can be bypassed without the reader becoming lost in later explanations. The full physical details can be found in any standard optical textbook (Hecht 2015; Born and Wolf 1999).

2.1 INTRODUCTION TO THE PHYSICS OF OPTICAL MICROSCOPY

2.1.1 Light

All the instruments in this book use light as the source of energy with which to probe the sample. It is then subtle, or sometimes blatant, changes in the form of this energy, as it interacts with the sample, that subsequently appear as features in the image. The fact that light is a form of energy should not be forgotten as this conveys the important message that, as we observe a sample, we are perturbing it by delivering energy. Thus, to view the sample with minimal risk of changing its state the light level should be kept as low as possible. As a source of energy light forms part of the electromagnetic spectrum and is conventionally considered as being the portion that can be detected by human eyes. This is normally considered to cover the region from around 400 to 700 nm. In terms of colour this extends from the violet at the short wavelength end up to the deep red at 700 nm. Figure 2.1a illustrates the wavelength ranges covered by optical microscopes. At shorter wavelengths one has moved into the ultraviolet and at the longer wavelength end into the near infrared beyond 700 nm.

It is perhaps interesting to note that the most commonly used fluorophore, until chemical and detector developments over the last twenty years, is fluorescein. Apart from being highly water soluble and non-toxic, one of its main advantages is that its peak emission is in the green, close to the maximum sensitivity of the human eye. It was thus the fluorophore that could be most easily detected before the advent of sensitive electronic detectors. In this book we will consider optical microscopes to operate down to around 250 nm and up to around 1500 nm. At wavelengths longer than this water absorption can become an issue and one has to use increasingly complex detectors (Section 2.6).

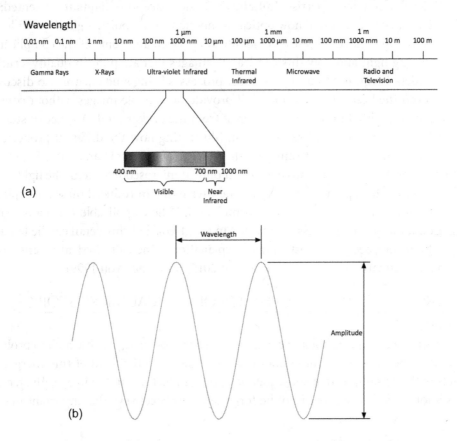

FIGURE 2.1   a) The region of the electromagnetic spectrum covered by optical microscopes discussed in this book. b) Defines the terms used for a wave.

2.1.2 Waves, Wavelength, Frequency and Particles

Since the middle of the 16th century there has been a controversy over the nature of light. Is it a wave (classical physics) or particle (quantum mechanics or Newton's "corpuscles")? Many of the most famous physics experiments undertaken with light involve demonstrating its wave-like nature or attempting to determine the speed of light, in both a vacuum and also in a material (gas, solid or liquid). We now know that light can be considered to exist in both "forms" and this plays a fundamental role in our current understanding and interpretation of quantum mechanics. As will become clear throughout this book, explanations are given that use both manifestations of light, and the one which is believed to provide the clearest description for a specific process will be used.

As a wave light obeys the equation $c = \upsilon\lambda$, where υ is the frequency of the light, λ the wavelength and c the speed of light in a vacuum. These features of a wave are illustrated in Figure 2.1b. In 1983 the numerical value for the speed of light was fixed at 299,792,458 ms^{-1} and the definition of the length, the meter, set from this value. According to Einstein's theory of special relativity c is the maximum speed at which light, and hence information, can travel.

The other fundamentally important equation expresses the energy present in light for an individual photon. In Planck's equation $E = h\upsilon$ where E is the energy of an individual photon, h is Planck's constant (6.63×10^{-34} Js) and υ again the frequency of the light. If one has a light source which is delivering 1 mW of light at 500 nm onto a sample one can therefore determine the number of photons that are actually arriving per second using $P = Nh\upsilon = \dfrac{Nhc}{\lambda}$ where P is the power, and N the number of photons. For the example given this would be 2.5×10^{15} photons per second.

As a wave, light has a) wavelength, the distance from one peak (or trough) on the wave to the next one, and b) frequency, the number of waves passing a point in a second. These are illustrated in Figure 2.1b. A single particle of light is known as a photon, and the more energetic a particle the shorter its wavelength.

2.1.3 Refractive Index

Light travels more slowly through any material than it does through a vacuum. In his corpuscular theory of light this was one of the predictions that Newton got wrong, as his theory (based upon light being corpuscles) predicted light would be faster in glass than in air. The ratio of the speed of light in a vacuum to that in a material is defined as the refractive index (n). For most gases the refractive index is close to 1 whereas in glass the refractive index is around 1.5 meaning light travels at a speed of around 2×10^8 ms^{-1}. However, this is not the whole story as the refractive index of a material is dependent on the wavelength of light. This effect can be both an advantage, for example in prisms where it is used to separate different wavelengths of light, and also a disadvantage leading to different wavelengths being focused to different points by a lens (see Section 2.4).

In a beam of light the way that one part of the wave travels relative to the rest of the beam is known as its *phase*, and the speed at which an individual crest or trough travels through a material is known as the *phase velocity*. If we have a light source with more than one wavelength, unless the light is travelling through a vacuum, one wavelength will have a higher phase velocity than the other. It is the phase velocity that provides the value used to determine the refractive index. In most cases we do not have a single wavelength (even from a laser there is a small spectral bandwidth) and thus some wavelengths will be slowed more than others. If we put in an ultra-short pulse of light, such as that used in non-linear imaging (Chapter 8) the light may have a wavelength range of over 20 nm and then the longer wavelength, redder light (which generally travels faster through a material) will move ahead of the shorter wavelength, bluer light. This process is known as dispersion and is the basis of a prism. In the case of our short pulse of light it means that the pulse of light gets stretched in time and the level of stretching depends on the material and the spectral bandwidth of the pulse. The *group velocity* is the term used to then describe the velocity of the entire pulse envelope.

As light travels at a different speed in materials, if light enters a material with different thicknesses across the aperture, part of the beam will take longer to emerge leading to a distortion of the wave. This is used both in lenses and also as the source of contrast in phase contrast microscopy.

2.1.4 Polarization

Polarization is the property of the direction of the wave's oscillation. For a wave travelling in the x direction the waves can either oscillate in the y or z directions, or a mixture of the two. In the case of light, which is a combination of oscillating electric and magnetic fields, it is the direction of the electric fields' oscillations which defines the direction of polarization. In Figure 2.2 the light is vertically polarized. A beam is *linearly polarized* if all the waves are oscillating in the same direction, and *un-polarized*, or *randomly polarized*, light is a combination of waves oscillating in all directions. Light that is un-polarized can be separated into two polarization states (vertical and horizontal) using several components including specially coated optics, certain crystals and a specific polarization selected using special plastics (such as that used in sunglasses).

The other form of polarization is that of *circularly polarized* light. This consists of two perpendicularly electromagnetic plane waves of equal amplitude with a 90° phase difference. If you were looking at the light as it came towards you, and were able to detect the polarization, you would see the direction of the polarization rotating. If this was moving in a clockwise direction then this is defined as *left-circularly polarized* light. The rotation of the light

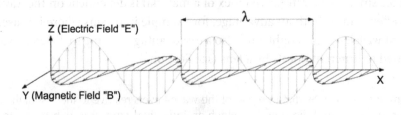

FIGURE 2.2 A representation of a polarized light wave.

would make one complete revolution as the light advanced towards you by one wavelength. Circularly polarized light can be converted to linearly polarized light by passing it through a quarter-wave plate, and vice versa. A half-wave plate will convert horizontally polarized light to vertically polarized light.

Light waves passing through, or reflecting off, materials can have their polarization altered, either rotating the angle of the polarization or increasing, or decreasing, the level of polarization. The scattering of light by particles can affect the polarization and the effect is dependent on the material and the size of the particle.

2.2 THE OPTICS OF MICROSCOPY

Having established the properties of a light wave these now need to be put into the context of an optical imaging instrument. It is the wave-like properties of light that set the resolution limits in optical microscopy. The main considerations are what happens to light when it passes into a new material and what happens when a light wave is spatially restricted.

2.2.1 Refraction

In the preceding section the concept of the refractive index of a material was introduced as a measure of the speed of light in the material relative to that in a vacuum. If the light wave hits the material at normal incidence then it will proceed into the material, just slowing down. However, if the light wave hits at an angle it is bent towards the line of the normal as illustrated in Figure 2.3a. As an analogy the light coming towards the interface can be thought of as a line of marching soldiers. As they approach some wet ground the first soldiers to enter will be slowed down meaning that as they march deeper the line will appear to swing in towards the end that has entered the wet ground first. As they leave

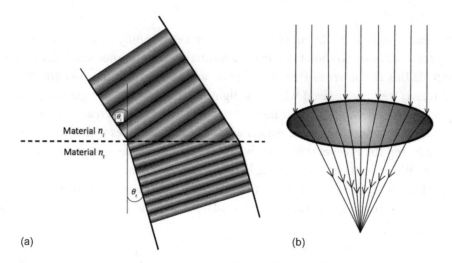

(a)   (b)

FIGURE 2.3   a) Light refracted at a plane surface showing it is refracted towards the normal. b) The use of refraction to focus light using a convex lens.

the reverse will happen and the first soldiers out will start to advance further in the same time it takes the others to leave the sticky ground. The net result is that the line of soldiers, or light beam, will be bent towards the normal in the material and then bend away as it leaves the block, therefore the incoming and outgoing rays will have the same angle.

The refraction of light as it enters a new material follows Snell's law, which was derived in the early 17th century. With reference to Figure 2.3a it can be proved by considering how far light travels in a given time in the two media linked with some simple geometry. This gives the well-known $n_i \sin\theta_i = n_t \sin\theta_t$ definition of Snell's law. On the exit from the block the angles and values of n are reversed and so the outgoing beam is parallel with the first, but it has been displaced. It thus appears to have come from a different entry point if the viewer does not know about the material's refractive index affecting the beam path.

A refractive lens uses the refractive index of glass and the angle at which that the light hits the surface to bring light to a focus. If the material has a curved surface as a bundle of light rays hit they will be bent at different angles, but all in the same direction, causing the light to come to a focus for a positively curved surface as shown in Figure 2.3b. For a concave surface the bundle of rays move away from each other leading to a diverging lens. The distance from the centre of the lens (for the bi-convex lens shown) to the focal point is known as the focal length of the lens.

The exact focal length of a lens is not trivial to calculate but if the lens is treated as having no thickness, a so called *thin lens*, then its focal length f is given by the "Lensmaker's Formula"

$$\frac{1}{f} = (n-1)\left(\frac{1}{R_1} - \frac{1}{R_2}\right)$$ For a typical glass lens with a refractive index of around 1.5 and 100 mm radius of curvature on one side and 200 mm on the other this gives a focal length of 400 mm. Care needs to be taken to ensure that the correct value of the radius is taken and that the correct definition of the direction of the radius is considered.

A full mathematical treatment of the way that an electromagnetic wave interfaces with a transparent surface also shows that even at normal incidence some small fraction of light is reflected. This is known as *Fresnel reflection* and depends on the two different refractive indices as well the angle at which the light hits the surface. For glass this provides a reflection of around 4% for light at normal incidence and this increases with angle and a higher refractive index. This is the reason why all optical surfaces within a microscope are coated, reducing the Fresnel reflection, which would otherwise give ghost images within the system. When the light is travelling between materials from a high to a low refractive index the level of reflection increases with the angle until a point is reached at which all the light is reflected. This is known as *total internal reflection* and is the physics behind optical fibres.

For light that is travelling from a material with a high refractive index (n_1) to one with a low refractive index (n_2) the critical angle at which light is first totally internally reflected is given by $\theta_{crit} = \arcsin\left(\dfrac{n_2}{n_1}\sin\theta_2\right)$ when $\theta_2 = 90°$. For a glass to air interface, with the glass having a refractive index of 1.5, the critical angle is 41.8°; for glass to water the angle is 62.4°.

2.2.2 Interference and Diffraction

Although they are normally introduced as two different effects there is no significant physical distinction between interference and diffraction. Waves are said to interfere destructively when the peak of one wave adds to the trough of another wave with the net result of zero (or no light). In the opposite case, when the peaks are matched, then constructive interference takes place giving a bright spot (Figure 2.4a). For such effects the light waves have to be in phase, meaning they have the same wavelength and that their movements are locked together (they go up and down collectively). Interference is an inherent property of all waves. The classic experiment to show light is a wave is perhaps Thomas Young's two-slit interference experiment.

Interference between waves is then the main physical effect leading to diffraction. It is diffraction that sets the classical limit in terms of spatial resolution within a microscope. This is best understood by considering how light passes through a hole. At the hole the light wave can be considered as a series of individual, very small oscillators each producing a wave of light, and all of the oscillators moving up and down together in phase (Figure 2.4b). As each of these "wavelets" moves out from the aperture the waves will start to interfere leading to a strong central peak, but with a series of concentric peaks, and blanks, away from the centre producing the classical circular diffraction pattern. The separation of the different rings is determined by the size of the hole and the wavelength of the light. The smaller the hole, the wider the separation of the rings, but the broader each ring individually is. A shorter wavelength of light will produce a smaller central region, as will be

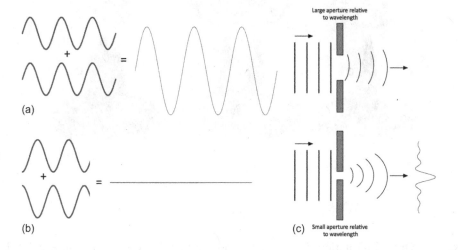

FIGURE 2.4 a) Constructive interference, b) destructive interference, c) diffraction through an aperture showing the smaller aperture causes greater diffraction or spreading of the waves.

discussed in Section 2.3. Diffraction at the edge of an optical system thus sets the ultimate spatial resolution, the "Abbe limit"

However, the Abbe limit is only achieved when all of the optics can be considered to be perfect, and this is not the case. We now consider some of the main aberrations that appear in nearly all optical systems, and crucially how the microscope manufacturer avoids these, but which can be very easy to reintroduce if the system is not used correctly.

As can be found in any basic textbook on optics or waves, for an aperture (and initially consider a slit) of width a the amplitude of the light exiting the aperture, $\varphi(u,v)$, is given by $\varphi(u,v) = a\,\mathrm{sinc}\left(\frac{au}{2}\right)\delta(v)$, where u is the direction through the aperture and v is in the direction of the height of the aperture. As it is the intensity of a wave that one sees, given by the amplitude squared, the intensity looking along the centre height of our aperture $v=0$ has a pattern given by $|\varphi(u,v)|^2 = a^2 \,\mathrm{sinc}^2\left(\frac{au}{2}\right)$. This is shown in Figure 2.4b. In the case of a circular aperture then the pattern is more complex and produces a circular pattern. The intensity of the pattern is determined by circular terms known as Bessel functions producing the classical set of concentric rings of intensity known as the Airy function or pattern.

The radius of the Airy disc is given by $r = 1.22\left(\dfrac{R\lambda}{2a}\right)$ where a is the radius of the aperture, λ is the wavelength of the light and R the distance from the centre of the aperture to the edge of the Airy disc (see Figure 2.5).

For a circular aperture the central peak is known as the *Airy disc* and its diameter is inversely proportional to the radius of the hole and proportional to the wavelength. This is important both for determining the resolution of an imaging system (as will be explained below) and also affects the settings in a confocal microscope as discussed in Chapter 5.

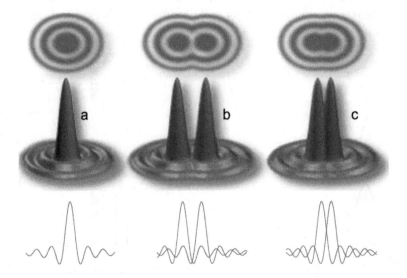

FIGURE 2.5 Optical resolution determined using the Rayleigh criteria, a) single point, b) two resolved points, c) two unresolved points. (3D plots credit to Molecular Expressions.com at Florida State University).

2.3 LIMITATIONS IN OPTICAL MICROSCOPY

We now need to consider the limitations placed upon optical microscopy, which are dominated by the wave-like nature of light. For readers interested, similar limitations can be found by considering light as particles and using the uncertainty principle in relation to the momentum of light (Padgett 2009; Stelzer and Grill 2000). But how does diffraction limit the spatial resolution when using a microscope to observe a sample?

Initially let us consider that we are looking at a sample, in which two point sources are emitting light into the optics of the system. They may be two slightly separated features in the sample. Each point will produce an image on the detector of the microscope, be it an observer's eye or camera. Each of these images will consist of an independent Airy pattern and as the point sources in the sample become closer together the two Airy patterns will start to overlap and at some position we will no longer be able to separate the two spots in the image. We have thus tried to image something beyond the system's resolution. The intensity profile of two such spots is shown in Figure 2.5. Lord Rayleigh stated that the two spots could be resolved (i.e. separated) when the centre of one Airy disc falls on the first minimum of the other. Although there are more mathematical routes to determining resolution this provides a very convenient definition (which was exactly the reason why Lord Rayleigh suggested it!). This minimal separation is proportional to the wavelength and inversely proportional to the diameter of the observing lens. This is linked directly with the numerical aperture (NA) of a lens which is given by $NA = n_i \sin\theta$ where n_i is the local refractive index and $\sin\theta$ is the half angle of the rays entering, or leaving, the lens (and hence related to the diameter of the lens and its focal length). This leads to the so called "Abbe resolution limit" where the smallest feature that a microscope can resolve is given by $d_{min} = \lambda/2NA$ (Abbe 1873). The basis for this formula is expanded upon in Chapter 12 on super-resolution microscopy where various approaches have been taken to overcome this limit.

In order to examine the optical properties of systems and to then model improvements one needs a method to quantify the deviations from perfection, or indeed even to know how "perfect optics" might perform. There are several systems but the most commonly used are those developed by Fritz von Zernike. He introduced a sequence of polynomials that are orthogonal and each one of which can be associated with a specific form of aberration. Orthogonal here means that if we adjust one term it has no effect on the other terms, thus we can separate out each aberration. The other important feature of Zernike polynomials is that they are orthogonal over a circular disc, which is the shape of many optical elements. One can thus look at the wavefront (phase) of a light beam and convert the pattern into a series of terms to express its deviation from a flat uniform wavefront using the Zernike terms. The full mathematics is beyond the scope of this chapter and can be found in any standard graduate level optics book. The important point to note is that mathematically there are two subscripts used to define each term (normally expressed as n and m) and that optically these are given one value which relates to the so called Zernike term which is then associated with a specific form of aberration. These are listed for the first few terms in Table 2.1.

TABLE 2.1 The First 10 Zernike Terms and
Their Associated Aberration Description

n	m	Z	Aberration Term
0	0	0	Piston
1	1	1	Tilt X
1	−1	2	Tilt Y
2	0	3	Defocus
2	2	4	Astigmatism at 0°
2	−2	5	Astigmatism at 45°
3	1	6	Coma in X
3	−1	7	Coma in Y
3	3	8	Trefoil at 30°
3	−3	9	Trefoil at 0°
4	0	10	Spherical

2.3.1 Chromatic Aberration

As discussed in Section 2.1.3 the exact refractive index of a material varies with the wavelength of light. This means that in the case of a simple lens the different wavelengths of light are refracted (or bent) by a different amount, meaning that they come to a focus at a different distance from the lens. This is illustrated in Figure 2.6a. In general, the shorter blue wavelengths of light are refracted more than longer wavelengths meaning that they come to a focus closer to the lens. If one then images an object in white light the images will be slightly blurred with each wavelength of light producing a slightly different image. This can be seen with a colour camera or one's eye as colour "halos" around an object when viewed in white light. In order to overcome this so-called *chromatic aberration*, optical systems are designed with multiple lenses where different refractive index materials are used to balance out the effects of this distortion through a combination of glasses and curvatures. In Chapter 3 the details of different microscope objectives (objective lenses) and the wavelength range over which they are chromatically compensated are discussed. Reflective optics, e.g. curved mirrors, do not produce chromatic aberration, as the light does not enter the mirror material.

2.3.2 Spherical Aberration

The natural shape of a polished lens is a section of a sphere. However, this shape leads to an effect known as *spherical aberration*. As illustrated in Figure 2.6b, for a parallel beam of light entering a lens the light rays in the central region focus further away from the lens than those entering at the edge. This is significantly more pronounced for optics with a steep radius of curvature and the effect is present in both refractive and reflective optical elements. As a rule of thumb the effect becomes significantly larger when light enters the lens beyond around two-thirds of the diameter of the lens. Thus, one solution is to have larger lenses and then to only part-fill them with light, though clearly this is not an efficient way of building an optical system. Again, through careful design and effectively distributing the focusing power across several surfaces the effect can be reduced. A further

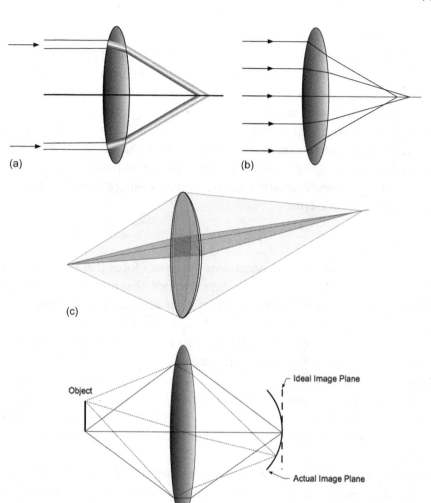

(a)

(b)

(c)

Object

Ideal Image Plane

Actual Image Plane

(d)

FIGURE 2.6 Aberrations in an optical system: a) chromatic aberration, with shorter wavelength light coming to a focus closer to the lens than longer wavelengths, b) spherical aberration: more peripheral rays focused closer to the lens, c) astigmatism: vertical and horizontal planes focus at different points along the optical axis, d) field curvature: focal plane is not flat leading to blurring at the edge of the image.

alternative is to use aspheric surfaces though these are significantly harder to fabricate. A simple way to observe spherical aberration is in a cup of tea. The cup acts like a large reflective lens to produce a focus of light from a distant source (such as a light from across the room). This focus can be seen on the surface of the tea as a line stretching in the direction of the light source. Spherical aberration can also be caused in optical microscopy if one uses the wrong oil with an oil immersion lens, or images at depth through a sample containing water whilst using an air or oil objective. As will be discussed in Chapter 3, lenses are available which compensate for different refractive index samples using a correction collar. This effect can also be seen when the wrong thickness of coverslip is used and again is discussed in detail in Chapter 3.

2.3.3 Astigmatism

This is an effect in which the optical system has a different optical power or focusing ability in two perpendicular planes. This is illustrated in Figure 2.6c where the lens has a stronger focal power in the x axis than in y and hence light comes to a focus closer to the lens for horizontal features compared to vertical ones. The ultimate limit of this is in a cylindrical lens, which if illuminated with a beam of light produces a line focus. This effect is used in the selective plane illumination microscope (SPIM) to produce a sheet of light. Clearly this is not wanted in most imaging systems and can be present in both refractive and reflective optics. In a person's vision astigmatism can be corrected using a lens with optical power in the angle of the astigmatic error.

2.3.4 Field Curvature

This effect means that the optical system does not have a focus that is flat across the focal plane. This is illustrated in Figure 2.6d where the light that produces the image and the edge of the field of view has a slightly different focus compared to the light in the centre. This means that the image can be adjusted to be in focus at the centre but will then be out of focus at the edges. This again is partly a result of using spherical surfaces and is present in both refractive and reflective optical systems. The effect can be minimized by again restricting the aperture used (as in the case of spherical aberrations) or by adjusting the shape of the lens surfaces to compensate. A third option, but which is not normally practical in microscopes, is to use a detector that is curved and matched to the optical curvature! In Chapter 3 the level of correction for field curvature is discussed for different types of objective lenses.

Although these are not the only aberrations, or methods of quantifying aberrations, in an optical system they are the main sources of poor-quality images. Significant research and design effort has taken place to minimize the effects of such aberrations in a microscope. One of the reasons for the high cost of many objectives designed to minimize aberrations, is that they can have 16 to 20 separate lens elements inside.

2.4 CONTRAST MECHANISMS

Beyond the optics of a system an equally important factor is the contrast present between the different features within a sample. In the simplest terms this is the ability for the eye, or computer, to distinguish between the two points. For example, a black dot on a black background cannot be seen, but if we place a local white ring or background around the dot then the dot will be rendered visible. In microscopy much of the skill of the experimenter is in determining how to maximize this contrast to enable the features of interest to be visualized against the background, for example the use of specific labelling and spectral filters to separate the various features. To some extent all developments in microscopes for the last 100 years have been attempts to increase image contrast. The following provides some background into the most commonly used physical phenomena as contrast mechanisms in optical microscopy. The methods by which these techniques are then exploited in a specific imaging modality are then expanded upon in the following chapters. It should be remembered that a physical process that may be used to maximize the contrast in one method may well act as a hindrance in another. For example, absorption of light can be a very powerful tool, but if the absorption is of the fluorescence generated within the sample, fewer photons will reach the detector resulting in a loss of contrast with the background. The other important point is that many of the processes are wavelength dependent and thus different processes may take place within the same sample depending on the wavelength of the illumination light.

2.4.1 Absorption

Probably the most common physical process used to obtain contrast in a microscope is that of absorption. Here the light is absorbed from the illumination beam and thus features appear dark on a light background. The absorbed energy has to go somewhere and may lead to heating in the sample or in some cases fluorescence as the energy is re-emitted at a longer wavelength. The absorption in a particular sample will be dependent on the molecules present and may occur over a narrow wavelength range or more broadly across the entire optical spectrum. For a specific absorber the quantity of light absorbed will depend on the number of molecules in the light path. This does mean the method, with careful control, can be used in a quantitative manner to determine the number of molecules in the optical path. However, the results might be due to a strong local concentration, or a column of molecules along the optical path. Clearly the wavelength illuminating the sample plays an important role in the level of absorption and again interesting information can be obtained by imaging at several selected wavelengths.

The absorption within the sample may be present due to naturally occurring compounds (for example melanocytes in skin) but can also be introduced through the addition of chemicals that preferentially label certain areas of the sample. Probably still the most widely used application of microscopy, in the life sciences, is in the area of histopathology where H and E (haematoxylin and eosin) staining is still the method of choice. In white light this leads to areas of the sample appearing purple in colour as some of the white light is absorbed on transmission through the sample.

2.4.2 Scattering

Scattering in a sample is used in two modes to produce contrast in the final image. Scattering occurs due to local differences in the refractive index of the sample. The level of scattering depends on (a) the local difference in refractive index, (b) the size of the particle or feature, and (c) the wavelength of the light. In a scattering process there is no change of wavelength with the exception of Raman scattering, which is discussed later in section 2.4.6. Light at longer wavelengths is scattered less than at shorter wavelengths which is one of the reasons near-infrared light is used for imaging through deep tissue samples. The exact way (level of scattering and angular range of scattering) that light scatters depends on the size of the particle relative to the wavelength of light, and different scattering regimes use different approximations to solve the complex mathematical equations which describe the process. At a larger scale scattering can be considered as being probabilistic and a value can be assigned to the change of a photon being scattered over a particular angular range.

The most common use of scattering in microscopy as a contrast mechanism is through the scattering of light out of the illumination path. This leads to darker areas within the image on a light background. Again, through the use of different wavelengths, and subsequent image processing, information on the size and nature of the scatterer can be obtained, though generally the method is not used in a quantified manner. In some samples, in particular those with fairly transparent fibre-like structures, colour "halos" can be seen around such features when white light is used for the illumination. This is due to the different scattering properties with wavelength.

An alternative use of scattering, and one not commonly used, is that of darkfield microscopy. As will be explained in Chapter 3, here the sample is illuminated obliquely so that under normal conditions no light reaches the detector. When a scattering sample is placed into the light path, at the focus of the instrument, photons can be scattered into the detecting optics to produce white features on a black background. Although the level of light scattered may be low as the initial background is dark, even a few photons reaching the detector present a high contrast feature on the final image thus providing a sensitive detection method for some samples.

Frequently scattering is a challenge in optical microscopy as in effect the scattering features deflect photons from their expected path leading to their detection in the "wrong position". In most of the three-dimensional, optical sectioning techniques described through the book it is scattering, of both the excitation and detection light, that limits the depth at which features can be seen. It should also be noted that the level of scattering is affected by the polarization of the light entering the sample relative to the orientation of the feature causing the scattering, and scattered light does have its polarization changed. Thus, through an examination of the polarization angle of the light further contrast enhancements are possible.

Beer–Lambert Law. In the case of both absorption and scattering a simple "law" is used to describe the level of light loss. This is known as the Beer–Lambert law, which relates the attenuation of light through a sample with the distance through the sample and the level of absorption or scattering present. Johann Lambert originally stated that the absorbance of a material is directly proportional to its thickness or optical path length. It was then stated by August Beer, in 1852, that the absorbance is proportional to the concentration. These two laws were then combined in order to quantify the loss of light through a sample to a concentration of absorbers (or scatterers) and the path length through the sample. The formulae stated below can be simply derived from considering the way that light travels through a sample, the distance l in the equation where there is a scattering, or absorption coefficient μ which depends on a scattering or absorption cross-section, and the number of such particles in a given volume, or the concentration. The intensity of light passing through the sample is then given by $I(l) = I_0 \exp^{(n_{scat}\mu_{scat} + n_{absorb}\mu_{absorb})l}$, where n_{scat} and n_{absorb} is the number density of scattering and absorbing particles respectively and μ_{scat} and μ_{absorb} the attenuation coefficients for the two effects respectively. When using these formulae (and there are multiple variations on this format of equations) take care in the correct use of units and also the exact definitions for n and μ.

2.4.3 Phase or Refractive Index Changes

As discussed above in Section 2.1.3 the refractive index of a material determines the speed at which light travels through a sample. Therefore, if a plane wave of light passes through a sample in which different areas of the beam travel through areas of different refractive index, the wavefront will no longer be uniform and parts will be delayed relative to others. This is known as the phase of the wavefront and cannot be seen on normal optical detectors, but can show areas in a sample which have different refractive indices but are otherwise transparent. The local refractive index variations may come about due to chemical concentrations, varying depth or the difference between water and lipids. All of these differences can therefore be used to create contrast in an image. In order to make the phase changes visible a variety of optical methods have been developed including phase contrast microscopy and DIC imaging. Each has advantages and drawbacks and these are discussed in Chapter 3. The use of phase-based contrast has shown decrease in use as fluorescent-based methods have become widely adopted, but as one of the so-called "label free" imaging methods (i.e., those which do not require the addition of a compound to develop the contrast in the image) the method should always be considered, in particular for single cell imaging.

2.4.4 Fluorescence

In one form or another fluorescence is probably the most widely used contrast mechanism after route H and E labelling. Here light of one wavelength is absorbed by a compound followed by light emission at a slightly longer wavelength. Through the use of dichromatic filters which can be used to separate the excitation from emission light it is thus possible to observe images of the fluorescing areas. Through the use of multiple fluorescent compounds it is possible to build up complex images where different features can be excited by different wavelengths. One advantage here is that through chemical manipulation the fluorophore can be targeted at specific features or areas of the samples to provide local and specific detail. In the case of cell biology a specific receptor or chemical, for example calcium, can be fluorescently labelled.

In the last twenty years the use of fluorescence has grown rapidly for both technical and biological reasons. With the improvement in camera-based detection systems (see Section 2.6) and specific wavelength sources (see Section 2.5), high-speed and high-resolution fluorescence imaging is possible with subsequent computer analysis of the data. In a biological context the use of genetically encoded fluorescent proteins within a very wide range of animals, plants and cells has transformed many imaging-based studies. The award of the 2008 Nobel Prize for this method is an illustration of the importance of the technique. Originally the fluorescent protein was excited in the blue and emitted in the green (green fluorescent protein or GFP which naturally occurs in the jellyfish *Aequorea victoria*) but now through chemical manipulation there are more than 100 fluorescent proteins that can be encoded covering the entire visible spectrum. As will be seen throughout the following chapters fluorescence is crucial as the contrast mechanism for nearly all three-dimensional and super-resolution microscopy.

2.4.5 Fluorescence Lifetime

Beyond the range of wavelengths that can be generated from a range of fluorophores described above there is one other feature that makes fluorescent-based methods very interesting. There is a short delay in the time from the absorption of the photon to the subsequent emission of light at a longer wavelength. This delay, typically of the order of nanoseconds for common fluorophores, is dependent on the local environment (pH, concentration, viscosity, etc.) and thus provides an additional contrast mechanism. With the advent of comparatively low-cost ultra-short pulse lasers and high-speed detectors and electronics, fluorescent lifetime imaging (or FLIM) is now a technique which is entering the mainstream methods of imaging. The full details on the processes involved in fluorescent lifetime imaging as a contrast mechanism are covered in Chapter 6.

2.4.6 Non-linear Excitation; Harmonic Generation, Raman Scattering

All the processes discussed above are described as being linear in that the signal detected (either loss of light or fluorescence) is, at least to a first approximation, linear with the excitation intensity. If you double the light going in, the contrast is also doubled as long as saturation has not been reached. However, with the advent of high-power, and crucially ultra-short pulse lasers (those typically with a pulse length of a few picoseconds (10^{-12} s) or shorter), effects which are not linearly related to the input have grown in importance. Here the high peak power and short timescale mean that the peak intensity of the light is very high and a number of effects can take place. The best known is that of two-photon excitation, in which two photons in the near infrared are absorbed simultaneously to excite fluorescence normally stimulated with a single photon at around half the wavelength. This technique is described in full in Chapter 8 and a further variation, in which the light emerging from the sample at exactly twice the frequency (half the wavelength) is detected, known as second harmonic imaging, is covered in Chapter 9.

The other contrast mechanism is the use of Raman scattering (either linearly or non-linearly excited). Here the light that emerges from a sample has its wavelength altered (normally to a longer wavelength) with specific quantities of energy being lost to molecular vibrations within the sample. The Raman process was described nearly 90 years ago, but it is only with recent technical developments and advanced computing methods to interpret the images that Raman microscopy has emerged as another of the label-free imaging methods. The full details are described in Chapter 10.

2.5 A BRIEF INTRODUCTION TO LIGHT SOURCES FOR MICROSCOPY

As described above, technical advances in both light sources and detectors have played a major role in the advances and plethora of microscopy methods that have recently emerged. Those currently available are described briefly below to provide the reader with a basis upon which to may make a decision when selecting a microscope for a specific task. In Chapter 3 the correct adjustment of the illumination source to obtain the best images (Köhler Illumination) is described and in the comments below it is assumed that all the sources have been correctly adjusted.

2.5.1 Conventional Filament Bulb (Including Metal Halide)

This is still the most common illumination source used in all microscopes as it provides a low cost and easy to control illumination. A metal filament is enclosed in a low-pressure atmosphere within a glass envelope and a low voltage passed through the filament. As one adjusts the current flowing through the filament it glows more brightly and also changes in colour becoming whiter (more blue light) at higher powers. In the case of a metal halide bulb the intensity is much higher and it is capable of producing ultraviolet light. Due to their high brightness metal halide bulbs can be used for fluorescence microscopy but more conventional tungsten bulbs do not have sufficient spectral brightness. Control of the intensity is straightforward but the changing colour of the light with brightness can cause problems in some cases. When this happens the intensity of the bulb can be adjusted with filters rather than direct adjustment of the intensity. The illumination system can become hot, especially in the case of the metal halide sources.

2.5.2 Mercury Vapour Bulb

Mercury vapour bulbs are the traditional route to producing a very intense light suitable for fluorescence microscopy. In these sources a gas discharge is excited within a quartz envelope containing low-pressure mercury vapour. This produces an intense white, and generally ultraviolet, light that results in local ozone production. Due to this ozone production, and the very high temperatures that bulbs can reach, the bulbs are not suitable for use in certain confined areas. Through careful control of the bulb pressure it is possible to have a bulb with high spectral brightness, thus these sources are used for fluorescence microscopy with excitation down into the ultraviolet for fluorophores such as DAPI. They are generally expensive and only have a limited lifetime (typically around 200 hours) when they should be replaced to minimize the risk of the bulb exploding and releasing mercury vapour.

2.5.3 Light Emitting Diodes (LEDs)

Since 1996 and the development of high-power blue and ultraviolet LEDs the use of LEDs as an excitation source for optical microscopy has grown. In particular the advent of high-power white LEDs (which are based upon a UV LED exciting white phosphors) are rapidly changing the way that conventional microscopy is undertaken. White LEDs are now powerful enough for most conventional forms of microscopy and when integrated with specific coloured LEDs can be used for fluorescence imaging. The major advantages of LEDs are their low cost, exceptionally long lifetimes (10,000s of hours), low power consumption and low operating temperature. In all cases, including white LEDs, altering their intensity does not affect the colour of the light and thus spectral brightness is maintained with changing intensities. At present they are probably the light source of choice for most microscopy "standard" applications.

2.5.4 Arc Lamp

Before the advent of lasers, arc lamps were the brightest light sources available. Crucially for microscopy the light can be produced in a very small volume and thus re-imaged into the sample for very high intensity illumination, even if the source is spectrally filtered. In all arc lamps a visible discharge in a gas is produced by striking an arc. The arc is then maintained through the current supplied. Generally the intensity of the arc cannot be altered as changing the power can cause intensity instabilities, thus the intensity is controlled by filters or apertures. The arc is normally instigated and controlled by an electrical discharge but lamps in which the arc is driven by a laser diode are now available with very high powers. Arc lamps offer very intense illumination but at a high cost in electrical energy, heat production and money! They also have a limited lifetime that needs to be monitored.

2.5.5 Lasers

Lasers have caused a revolution in optical microscopy and as will become clear throughout this book form the cornerstone of nearly all the advanced developments in the last thirty years in optical microscopy. Their main benefit in terms of optical microscopy is their narrow spectral width combined with high intensity. When these features are linked with recent all solid-state devices (based upon laser diodes) operating in the visible portion of the spectrum lasers become an interesting source. Due to their high intensity, and the ability to pass down single mode fibres efficiently, lasers are used in all beam scanned optical microscopy systems. The ability of lasers to produce very short light pulses also makes them the only source suitable for the wide range of non-linear microscopy methods described later, as well as fluorescent lifetime imaging. Due to the speckle effect within a laser beam (caused by the spatial coherence of the beam leading to local interference patterns) lasers are not the source of choice for wide field excitation. They offer a high spectral brightness, controllable source but generally at single wavelengths. Further details on the exact sources used for different forms of beam scanned microscopy are provided within later chapters. The features of each illumination system are summarized in Table 2.2.

2.6 DETECTION OF LIGHT IN MICROSCOPY

All the recent advances in optical microscopy have, at their heart, the use of advanced electronic detectors. When this electronic capture of the image is linked with the rapid growth in computer power one has the foundation of all the advances made in the last twenty years. Although lasers as an excitation source are important, in the end it is the efficient capture of the photons that produces the image that is crucial. Originally, it should be remembered, data was captured by the observer and recorded as a work of art.

2.6.1 Human Eye and Photographic Film

Clearly the most used detection system in microscopy around the world is the human eye. The main disadvantage with direct visual observation is the inability to quantify or record what is being seen, except through drawings, the main advantages clearly being ease of use and stereoscopic observation. It is probably this latter feature that is now the strongest driver for purchasing stereo, or binocular microscopes. Even in top end beam scanned systems the microscope is built around the conventional stereoscopic imaging optics and sometimes compromises are made in the beam scanning optics to maintain the

TABLE 2.2 Summary of Illumination Sources

Source	Advantages	Drawbacks	Microscopy Use
Filament Bulb	• Low cost • Easy to control intensity • Long history	• Colour change with intensity (or use filters) • Heat • Poor in UV • Not intensive enough for fluorescence	• Conventional
Metal Halide	• Very bright • Full spectral coverage	• Can be high price • Limited lifetime • Spectral change with intensity (or use filters) • Heat	• Conventional and fluorescence microscopy
Mercury Vapour	• Very intense • Excellent spectral coverage	• Heat + Ozone • Limited lifetime • Risk of explosion • Hg vapour • Hard to control	• Fluorescence
LED	• Low cost • Easy to control intensity and pulse • Good spectral brightness • Narrow spectral width and white light • Very long lifetime	• Multiple LEDs for good spectral coverage • White light from UV, with uneven spectrum	• White for conventional • Coloured fluorescence • Super-resolution
Arc Lamp	• Very intense • Full spectral coverage	• Heat + ozone generation • High cost • Physically large • Hard to control • Limited lifetime	• Fluorescence
Laser	• Easy to control intensity • High spectral brightness (monochromatic) • High quality beam profile • High intensity • Good spectral coverage	• Potential high cost • Monochromatic • Speckle • Not total spectral coverage	• Fluorescence, non-linear, Raman, super-resolution

conventional viewing method. Direct visual observation provides the user with a clear view of the sample and the ability to position items such as probes and electrode patch clamps in the correct place in three dimensions. The eye also inherently has a large dynamic range, meaning it can frequently detect features in a sample that can be missed by more technically sophisticated methods. The other feature of direct observation is that the eye is hard wired into a phenomenally powerful graphics processing unit in the brain. However, this system can also be tricked into seeing things that might not be present, as is known from the observation of optical illusions.

The use of the eye as the detector in optical microscopes still dominates the overall design of the system and has clearly played a significant role in the development of microscopy. An illustration of this can be seen in relation to fluorescence microscopy where for many years fluorescein was the dominant fluorophore used. While the fact it is non-toxic and highly water-soluble is clearly a factor in its wide application, fluorescein is also bright but perhaps most importantly it emits at the peak of the human eye's sensitivity, in the green. Thus it was, and still is, the fluorophore most likely to be seen through human observation.

Although virtually unused now, photographic film was a major detection system in microscopy as for decades it provided the only reliable method of recording an image. Special high-sensitivity films were developed for use in microscopy, and if one considers that each light sensitive molecule as a pixel, with a size of a few nanometers, film still provides the highest number of pixels in an image! Since the middle the 1990s, however, wide field detection has been dominated by electronic cameras.

2.6.2 Electronic or Digital Camera

With the development of the low-cost personal computer in the early 1980s, and the increasing performance of detector arrays, the electronic or digital camera is now the most widely used method of recording images in wide field microscopy. Figure 2.7 illustrates the basic physical principles behind a digital camera. In all such detectors a photon is absorbed into a material and an electron released. In a camera system this electron is then directed towards a collecting "well". The electronic contents of each well (typically one well per pixel) is subsequently "shifted" out to be read out sequentially on demand. The size of the charge present is digitized to determine the intensity on that pixel. In the digitization process the size of the charge is converted into a number of different levels, which determine the number of intensity bits in an image. Thus, for an 8-bit image the intensity is converted into 256 levels (zero being black and 255 being the maximum) or 1024 or 4096 levels for 10- and 12-bit images respectively. The higher the number of intensity bits the greater the intensity resolution.

The other figure given for cameras is the number of pixels that are used to form the image. Thus, for a camera which has 1000 by 1000 pixels the image will consist of 1 Mb image points. In order to obtain colour images each group of four pixels is covered by a "Bayer" filter. This is illustrated in Figure 2.7 inset. The Bayer filter consists of one blue filtered pixel, one red and two green to represent the relative spectral sensitivity of the human eye. The computer then processes these individual pixels, with a colour look-up

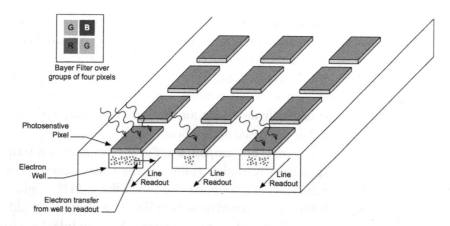

Bayer Filter over
groups of four pixels

Photosenstive
Pixel

Electron
Well

Line
Readout

Electron transfer
from well to readout

FIGURE 2.7 Illustration of the principles of a digital camera with a Bayer filter, to obtain colour images, shown in the inset.

table, to produce a full colour image. It should be noted that such colour images are frequently described as being 24-bit, but this means that each colour consists of an 8-bit depth intensity image.

The sensitivity of a camera to a specific wavelength is determined by the exact material composition of the detector. Most visible cameras use silicon as the main detector material, though this may be "doped" with other elements to control the exact spectral characteristics. Silicon-based detectors operate from the near UV through to around 1000 nm though the sensitivity falls off rapidly after the natural peak at around 850 nm. The conversion rate of a photon into an electron is known as the quantum efficiency and this will vary with wavelength. A wide range of more exotic materials is used for longer wavelength (near infrared) detection. The earliest detectors operating in this region were based on selenium or germanium but both have a low quantum efficiency (the conversion of photons into electrons with 1 being perfect) and also a high noise level due to thermally excited electrons being available at room temperature. Other materials now used include lead sulfide, indium antimony, indium gallium arsenide and for wavelengths greater than 2 microns mercury cadmium telluride.

A range of different electronic configurations and production methods are used in the manufacture of digital cameras and these are beyond the scope of this brief introduction. The original devices were known as charge coupled devices, but as semiconductor-processing methods have improved the market is now dominated by complementary metal-oxide-semiconductor (CMOS) devices. This change has been driven by cost and the mobile phone industry. Scientific CMOS (sCMOS) cameras use the same manufacturing methods, but the associated electronics to read out and digitize the signal are more complex and lower in noise. This enables faster readout, grouping of pixels (known as binning) or reading out lines or regions of the camera.

There are several sources for the noise that will appear on a digital image. Firstly there is the random release of an electron from the semiconductor material. This is determined by the exact structure of the semiconductor and also by the temperature. At higher temperatures the electrons within the material have a higher probability of being released and hence converted into an apparent photon arrival. This effect is exponential with temperature and more prevalent for devices optimized for imaging longer wavelengths (where the photons have less energy). This effect can clearly be reduced by cooling the sensor either using an electronic or "Peltier" cooler or water or, in the most extreme cases, liquid nitrogen though this is now normally only used for detectors operating well above 1500 nm in the near infrared and thus not based upon silicon.

The second source of noise is from any amplification of the signal as it is transferred from the "well" to the digitization electronics. Following the amplification there is the potential error in the digitization and also the noise that can be found in the digitizer electronics. Once the signal is digitized additional noise is unlikely. However, it should always be remembered that the photons from the source themselves have an inherent statistical variation that follows a Poissionian distribution. Thus, one has to balance up the different sources of noise for a specific application.

At low light levels the electronic shutter on the camera is likely to be open for longer (allowing the well to fill with more electrons) but this can increase the number of thermal electrons as well as those produced by photon absorptions. The digitization noise is generally constant for each exposure and the electronic noise added, before digitization, will be determined by the level of gain used in the amplifier.

When selecting a camera for a specific application the different sources of noise, along with considerations such as the speed of imaging, will determine which is the optimal camera to use. The performance of cameras is accelerating very rapidly and thus readers are advised to look at the latest specification of cameras to determine the most suitable for their application, but the text above should provide some pointers on what to consider.

2.6.3 Photomultiplier

Photomultipliers are generally the most sensitive detectors used in microscopy and capable of detecting single photons. Until recently with the development of advanced semiconductor devices (described in Section 2.6.4), they were the detector of choice in all beam scanned microscopes. Generally, photomultipliers are produced as single detectors and are not available as arrays though there are a few very expensive exceptions.

The basic operation of a photomultiplier is shown in Figure 2.8. Each has a window at the front to permit the passage of the photons to the detector. A window is required as the device operates in a vacuum and generally the casing is made of glass. The photon then hits the photocathode and an electron is emitted. This electron is then accelerated by a positive electric field towards a second plate where the photon, now at a higher energy, hits the plate emitting several electrons. These are then attracted to a third plate (all known as a dynode) where each electron is again multiplied, and this cascade process takes place down a chain of dynodes before being detected by the anode. Thus, from a single electron emission at the photocathode a large number (typically over one million) electrons are detected at the end. One now has a larger current, though only at best in the micro-amp region, which can be converted into a trace on a screen or digitized for storage in a computer.

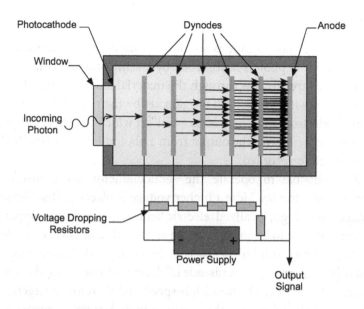

FIGURE 2.8   Basic photomultiplier configuration with five multiplying dynodes.

The exact material of the photocathode determines the spectral sensitivity and this can extend down to detect light below 200 nm (not typically used in optical microscopy) and all the way up to several microns. The quantum efficiency of a photomultiplier is defined as being the average number of electrons released, at the photocathode, for a single photon entering the device. The exact gain of a photomultiplier is determined by the voltage applied between the photocathode and the anode and the number of dynodes within the chain. Typical operating voltages are over 1000 V across the full dynode chain. The main disadvantages of photomultipliers is in the glass construction with the vacuum tube making them fragile, and the high operating voltages using a safety hazard, though they can be very sensitive. They range in size with the smallest being around 10 mm in diameter and about 50 mm long up to detection windows of 100 mm and lengths of 200 mm for collecting light from a widely emitting source. Photomultipliers can also operate at high temporal resolution for fluorescence lifetime imaging (see Chapter 6).

The main source of noise in a photomultiplier is the random emission of an electron from the photocathode and the fact that the multiplication is not the same for each captured photon. For this reason, the detection electronics often operate to "threshold" the signal so that only values above a certain voltage are determined to be caused by photons arriving at the cathode. The number of pulses above this threshold can then be counted in the so-called "photon counting mode". The level of noise, and variation of the signal, is strongly determined by the temperature, as with the digital camera and also by the accelerating voltage used along the dynode chain. Various methods of electronic enhancement are possible after the signal has been taken from the photocathode to improve the signal to noise including phase-sensitive detection for an intensity modulated source (including a pulsed laser for example).

2.6.4 Photodiodes

The operation of a basic photodiode is similar to that described above in the introduction to digital cameras. A photon is absorbed by a semiconductor material, releasing an electron, which is then free to flow through the material and collected at the anode of the device. The sensitivity with wavelength is again set by the material. As each photon produces only a single electron, and there is no "well" in which to collect the photons, a basic photodiode is not very sensitive. The output from a photodiode is normally then amplified before being displayed or recorded.

However, it is possible to operate the semiconductor using methods known as "avalanche" or "Geiger" modes. Here a higher voltage is placed across the semiconductor device to produce a very high, localized, electric field. The voltage, in comparison to a photomultiplier, does not have to be so high, as the semiconductor is very thin leading to very high fields (volts per mm) with only a few volts being applied. This means that if an electron is emitted then this can trigger a cascade of electrons to be released as the first electron accelerates towards the anode. The now high-speed and therefore energetic electrons will cause other electrons to be released as the energy of motion supplies energy to the electrons still bound within the semiconductor. One thus has, in effect, a solid-state photomultiplier, which can be small and operate at a low voltage. The disadvantage is that again single thermally excited electrons within the semiconductor can cause a similar cascade breakdown leading to false signals.

In the last ten years such avalanche devices have become significantly more useful and sensitive through the use of modern semiconductor manufacturing methods. Using the CMOS production process (used for nearly all semiconductor electronics) micro-structures can be created around a small area on the photodiode. This can lead to extremely high local electric fields and a localized avalanche of electrons on absorption of a photon. These devices are known as single photon avalanche photodiodes or SPADs (Gyongy et al. 2016). Due to the production process such SPADs can easily be assembled into arrays to produce in effect highly sensitive cameras. The SPAD pixels cannot be as small as on a camera as they require local structures to limit the high electric fields, but local, "on chip" electronics can be built into the device. Recent advances here have been to build in the electronics required for single photon counting to enable fluorescence lifetime imaging from multiple points. As with all of the "photon to electron" detection methods the noise in these devices is caused by the random release of an electron by the semiconductor and also slight variations in gain for each pixel in such a sensor. The manufacturing process for these SPADs is based upon standard electronic industry techniques and equipment, thus it is likely that such devices will become the detectors of choice for many beam scanned microscopy methods.

REFERENCES

Abbe, E. 1873. "Beiträage zur Theorie des Mikroskops und der Mikroskopischen Wahrnehmung". *Archives Microscope Anat 9*: 413–18.

Born, M. and W., Emil. 1999. *Principles of Optics*. 7th edn. Cambridge University Press.

Gyongy, I., N. Calder, A. Davies, N. A. W. Dutton, P. Dalgarno, R. Duncan, C. Rickman and R. K. Henderson. 2016. "SPAD Image Sensor for Microscopy Applications". In *IEEE International Electronic Devices Meeting*, 3–6.

Hecht, E. 2015. *Optics*. 5th edn. Pearson.

Padgett, M. 2009. "On the Focussing of Light, as Limited by the Uncertainty Principle". *Journal of Modern Optics* 55(18): 3083–9.

Pedretti, Ettore et al. 2018. "High-Speed Dual Color Fluorescence Lifetime Endomicroscopy for Highly-Multiplexed Pulmonary Diagnostic Applications and Detection of Labeled Bacteria". *Biomedical Optics Express* 10(1): 181–95. http://arxiv.org/abs/1809.01269.

Stelzer, E. H. K. and S. Grill. 2000. "The Uncertainty Principle Applied to Estimate Focal Spot Dimensions". *Optics Communications* 173(January): 51–6.

CHAPTER 3

Basic Microscope Optics

3.1 INTRODUCTION

This chapter covers the basic optics of a conventional widefield microscope, considering the collection of the illumination light through to the detection of the image by eye or electronically. Within each section some guidance is provided in the optimization of the microscope components along with some selection guidelines for specific applications. The aim of the chapter is to provide a core understanding of the main features of an optical microscope and how the various components work with each other to produce an optimal image for a specific application. Each section of this chapter can be read in isolation if the reader is interested in a specific area.

3.2 BASIC TYPES OF WIDEFIELD OPTICAL MICROSCOPE

Standard widefield microscopes are initially divided into two classes, inverted and upright. In an upright microscope the objective is mounted above the sample looking down at the features, whereas for an inverted microscope the objective lens is under the sample as illustrated in Figure 3.1. The selection of either an inverted or upright system is one that is based upon the samples to be imaged. In life science laboratories, inverted microscopes are generally more common as the sample can be placed on a thin glass slide and viewed through the slide. This provides access to the sample from above enabling chemical solutions to be changed and fluorophores added, and the ability to manipulate the sample. It is easy to reach the sample with a micropipette for patch clamp experiments in electro-physiology for example. As many biological samples are mounted in water an inverted configuration means that the lens does not have to look through the water medium. This means that one can directly view the cells, which are attached to the slide. Generally, it is easier to use higher numerical aperture oil objectives with an inverted configuration. Inverted systems normally have turrets in which multiple lenses can be mounted to enable easy changes of magnification.

Upright microscopes are generally more common in physical science or engineering laboratories. Here the sample is often air mounted and viewing from above, particularly in reflection microscopy, is therefore easier as one does not need to image through a glass coverslip. However, with the advent of higher numerical aperture water dipping lenses with

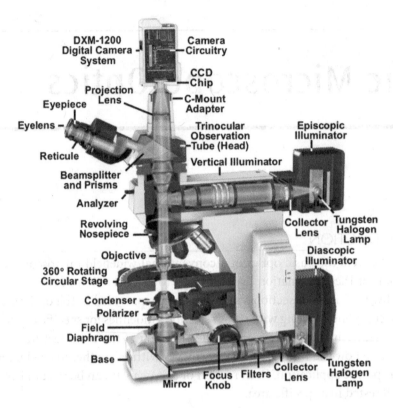

FIGURE 3.1   Basic optical components in an inverted microscope (from data sheet, by permission of Nikon Instruments).

long working distances, upright microscopes are growing in popularity in the life sciences. As the core optical system for a multiphoton microscope (Chapter 8) where imaging is frequently done *in vivo* the upright configuration is preferred as it makes sample mounting easier and gravity has less of an effect in squashing the features of interest against any glass coverslip. In the more expensive commercial microscopes the sample stage is fixed in the axial direction and all of the focusing is achieved by moving the objective lens.

A further consideration for a microscope configuration is based on the type of imaging to be undertaken. Again, in the physical sciences, in particular metallurgy or materials science, many samples are not transparent and thus the sample is viewed in reflection. Here the illumination comes through the objective lens in what is known as the "epi" configuration. It should be noted the epi configuration is most commonly used in fluorescence microscopy described in Chapter 4. The detected signal then passes back through the objective to be observed and recorded. The more common configuration in the life sciences, and for samples which have some level of transparency, is to use a transmission illumination system (also known as trans-illumination). Here the light is directed onto the sample on the far side of the objective lens, passes through the sample and then into the objective before being sent to the detector. This generally makes the basic optics mechanically easier to align and increases the light level that can be delivered to the sample. In trans-illumination one does not need a beam splitter in the path between the objective and detector.

The next division in the core microscope is between monocular and stereo-microscopes. The distinction is clear with monocular systems having one eyepiece and binocular or stereo-microscopes having two. A binocular system is clearly more complex but provides the observer with a level of depth perception making manipulation of samples, or patch clamping, easier. The clear choice, if the funds are available, is for a binocular system, but it should be remembered that if the main use of the microscope is to record images on a camera, those images will only be monocular and thus the stereo perspective offered through the eyepieces may be a luxury.

The initial detection of the image is nearly always made using the observer's eyes. Indeed it is really only in the last 100 years that images could be recorded on a camera, initially on film and then more recently on digital cameras. The human eye has a very large dynamic range and is very sensitive to low light levels across a broad spectral range. It is also much easier to manipulate the sample and find the correct area for imaging using one's eyes. This is because of the human brain's ability to provide feedback and correlation between physical movements and what the eye is seeing. Although moving a sample while looking at a screen displaying a camera image is clearly possible, it is less intuitive.

The final aspect of basic microscope optics and configuration only appeared around twenty-five years ago with the development of so called "infinity optics". The difference between conventional and infinity optics is illustrated in Figure 3.2. In the upper image (Figure 3.2a) it can be seen that the light emerging from the point of interest in the sample collected at the focus of the objective is collimated (parallel beams). This collimated beam

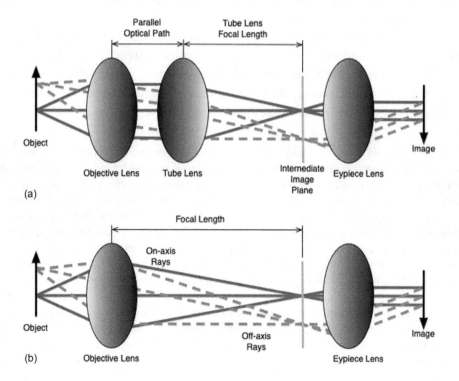

FIGURE 3.2   Optical diagram comparing a) infinity corrected optics, b) finite (traditional) optics.

will then pass through the system and can then be imaged onto the eye or camera using a tube lens and eyepiece. In Figure 3.2b the traditional, or finite, objective lens system is shown. Here the light emerging from the objective comes to a focus around 160 mm from the back of the objective lens. This focal (image) plane is then re-imaged onto the eye using the eyepiece. Clearly an infinity designed objective will not work in a finite microscope and *vice versa*. The main advantage of the infinity design is that extra optical elements (beam splitters, dichromatic mirrors, phase plates, etc.) can be placed into a collimated beam path, which is the best optical configuration. If a block of glass (beam splitter) is placed into a finite optical system it will move the back (i.e. rear) focal point as the glass will increase the optical path travelled by the light. All modern commercial microscopes now use the infinity corrected optical path and further details on the tube lenses used in combination with different manufacturers objective lenses are provided in Sections 3.5 and 3.6 of this chapter.

3.3 CORE OPTICS OF A WIDEFIELD MICROSCOPE

Figure 3.1 shows the major optical components of an inverted commercial microscope. All systems follow a very similar configuration and the same terms are used in both inverted and upright microscopes, and also transmission and reflective systems. The details of each section of the optics are described in the following sections.

The source of light is initially collected using a lens and then directed through an aperture (diaphragm) and onto the sample through a condensing lens. In the case of a system using epi-illumination (reflection or fluorescence microscope) the light is also collected, sent through an aperture and then directed onto the sample through the objective lens via a beam splitter or partially reflecting mirror.

The light, having either passed through the sample, or been reflected, is then collected by the objective lens and directed up through the optical system. In most modern microscopes there is either a movable mirror, or partially reflecting system, which sends the light to the eyepieces or onto a camera. The most common configuration is a movable mirror so that 100% of the light goes either to the eyes, or the camera.

Within a microscope's optical system there are two sets of conjugate planes. In Figure 3.3 the conjugate planes are illustrated showing these planes are present in both the illumination and detection optical paths. When the microscope is operating normally the conjugate set of object (or field) planes are all simultaneously in focus with the sample, thus if something is introduced into one of these planes (such as a gird or aperture) this will also be in focus. The other set of conjugate planes, aperture or diffraction planes, are in focus with the back aperture of the objective and can thus be seen if the observer focuses on this feature. Generally speaking the planes alternate through the optical system, and knowledge and understanding of the position and role of these planes is important for anyone adding features to an optical microscope such as DIC components, polarizers and filters. In simple terms if one wishes something to be in focus at the same time as the sample (graticule) it should be placed in an image plane. If, however, one requires the component to be out of focus (dichromatic filter so dust present does not appear in focus on the sample for example) then this should be in the illumination set of planes.

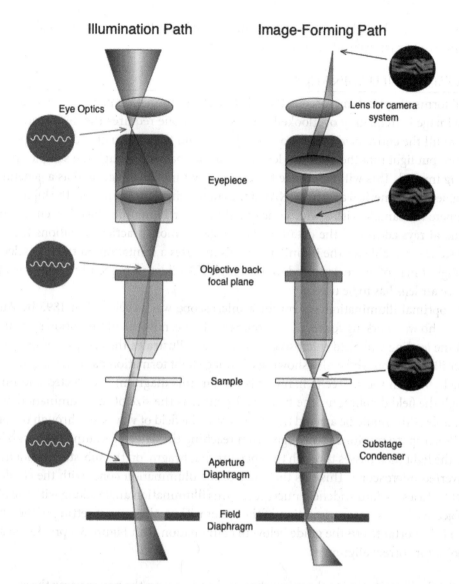

FIGURE 3.3 Köhler illumination.

For obvious reasons all of the optical components in the system should be designed to be as achromatic as possible as typical optical microscopes now operate from around 350 to 850 nm. Generally, most lenses are not simple, single element components, but compound lenses designed to minimize aberrations within the system. In modern microscopes, as will be specifically illustrated in the sections below, components from one microscope manufacturer cannot be readily interchanged with those from another, even if they are the same physical dimensions and fit into the system. Different designers compensate for optical distortion in different ways and even the focal length of components can be different leading to incorrect magnification for example. It is also obvious that components should be kept as clean and scratch-free as possible. Small particles of dust, scratches or finger marks can cause the light to be scattered, and hence lower the contrast on the final image, or in the worst case even add a feature to the sample!

3.4 OPTIMAL ILLUMINATION

For all forms of optical microscopy the role that the illumination plays in the final quality of the image is frequently overlooked. In basic terms one requires the illumination to be even, to fill the entire field of view of an objective and to come in at such an angle that it does not put light into the objective lens that has not passed through the sample area that is being imaged. This will maximize the contrast within the image, and as a general rule the higher the numerical aperture (NA) of the condensing lens the better. The high NA will illuminate the sample with a wide cone of light thus enabling the objective lens to collect peripheral rays adding to the detail in the image. In most practical situations the working distance available for the illumination lens imposes a limitation on the lens selection. Although it may often be preferred to use a high NA oil lens, frequently a longer working distance air lens has to be used.

The optimal illumination system for a microscope was developed in 1893 by August Köhler who was working for the Zeiss company. This is now the illumination system used by all the leading widefield microscopes. Figure 3.3 illustrates the critical components in Köhler illumination whilst also showing the image light formation path on the same scale.

The light from the source (a filament is used in this diagram) is collected and imaged through the field diaphragm. The field diaphragm sets the size of the illumination field at the sample. This should be around the same size as the field of view seen through the objective. This helps to minimize stray light from reaching the detector leading to a loss of contrast. The light then passes through the aperture diaphragm, on the sub-stage condenser in an inverted microscope. This sets the angle of the illumination cone (with the condenser lens NA). In a standard widefield microscope this illumination angle, along with the NA of the objective lens, sets the resolution of the system. Thus, the correct setting of the illumination is important, and the guide below in combination with Figure 3.4 provides a starting point for correct alignment.

1. Set the illumination to the minimum level and turn on the power using the switch on the illumination source. Switching a bulb on and off at the lowest setting helps to prolong its life as well as ensuring that the user does not subject their eyes to very bright light, which although not harmful can affect the eyes' sensitivity for several minutes.

2. Using a lower power objective (×10) focus on a thin specimen (flat lens cleaning tissue for instance) and close the field diaphragm until the edges can be seen in the image (Figure 3.4a).

3. Focus the condenser using the control under the microscope stage until the diaphragm is in sharp focus (Figure 3.4b).

4. Centre the image of the diaphragm in the field of view using the controls under the microscope stage. Opening the diaphragm can help with ensuring the diaphragm is as close to the centre as possible (Figure 3.4 c, d).

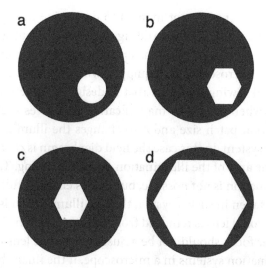

FIGURE 3.4 Position of the illumination mask during the adjustment of Köhler illumination, a–d illustrate the image seen at different stages in the alignment process.

5. Now fully open the field diaphragm.

6. Fully open the condenser diaphragm and then close slowly until the image contrast increases.

7. Remove eyepieces and observe the aperture diaphragm with even illumination arising from the ground glass screen in some microscopes, or in others, the filament may be seen.

8. The diaphragm is likely to be about 2/3 to 3/4 open (best compromise resolution vs contrast).

9. Replace eyepiece and observe.

10. For a different objective the process should be repeated adjusting both apertures. For example, for a ×40 lens after a ×10 objective close the field diaphragm and open the aperture diaphragm; but this may be a calibration on some microscopes on the aperture controls.

11. Adjust the intensity with the illumination control knob on the power supply.

12. **Do not** adjust the intensity by adjusting condenser lens or aperture diaphragm.

These instructions are written for a Nikon inverted microscope but are applicable to all systems with Köhler illumination. If this method is adopted one has an excellent starting point for high quality imaging. The intensity of illumination should be partially controlled using the lamp power supply, depending on the illumination source. In the case of an LED this is straightforward as once it is on only the intensity changes with electrical adjustment. To ensure the maximum performance of a tungsten-halogen bulb, however, one should refer to the manufacturer's instrument manual to determine the optimum lamp voltage (usually 6–10 volts) and use this setting. Adding or removing neutral density filters in the illumination path can then easily control the brightness of the illumination.

Some newer inverted microscopes are equipped with a specialized sub-stage condenser which has an additional "swing out" lens that is designed to be removed from the illumination optical path when using low magnification objectives such as ×2 and ×5. This increases the illumination patch size and also changes the illumination angles available through the condenser system. In this case the field diaphragm is now no longer as effective in determining both the area of the illumination and the centring. Therefore, without this lens true Köhler illumination is not possible, but it does ensure a full field of illumination. Care should always be taken in such a system that the illumination is aligned as described above, and then the optional lens is removed from the optical train. The system is designed such that the condenser focus should not be adjusted when the lens is removed. As a final comment on all illumination systems in a microscope, if the image being viewed is not as good as one expects it is probably worth checking the illumination path early on in trying to diagnose the fault. This is frequently overlooked as there is a general feeling that "if one can see something the illumination must be OK," and this is often a bad assumption!

3.5 MICROSCOPE OBJECTIVES

Probably the most crucial component in the microscope is the objective lens and financially they can form up to 50% of the total cost of an instrument. The aim of the objective lens is to collect as much light from the sample as possible. The fine detail is contained in the peripheral light rays from the sample and therefore in simple terms the wider the collection angle of the lens the higher the resolution. As was mentioned in Chapter 2 the resolution of a diffraction limited microscope is given by

$$d_{\min} = \frac{\lambda}{2n(\operatorname{Sin}\theta)} = \frac{\lambda}{2NA}$$

Where λ is the detection wavelength, n the refractive index between the lens and the sample and NA the numerical aperture of the lens. Thus to resolve the smallest features one requires short wavelength light, a high refractive index between the sample and the lens and a lens with a large numerical aperture. It is worth remembering that at this level the magnification of the lens is not relevant (though higher NA lenses tend to have higher magnifications). The wavelength of light is normally determined by the imaging and detector requirements, but n can be maximized through the use of immersion oil, and the NA through lens selection. It should also be noted at this point that the brightness of the collected light depends on the numerical aperture squared. However, this gain is sometimes lost as the magnification does play a role in the image brightness, which is inversely proportional to the square of the lens magnification.

In normal circumstances the half collection angle cannot be greater than 90° in air (due to the refractive index values of air and glass) but by using water rather than air between the sample and the first lens element, numerical apertures of up to 1.2 are possible, and with oil replacing water 1.5 NA lenses are available. Thus the use of higher refractive index materials helps to increase the NA as well as appearing in the resolution equation.

Objective lenses are described by the level of aberrations that are corrected in the optical design. The main aberrations, previously described in Chapter 2 Section 2.3, that are removed are spherical aberration, followed by chromatic aberration, at an increasingly wide spectral range, and then field flatness. As discussed in Chapter 2 all of the aberrations can in theory be removed through careful lens design, and Table 3.1 outlines the way that different lenses are corrected. It should also be noted here that objectives are designed to work either with, or without a cover slip, normally optimized for #1.5 coverslips (~175 μm) and for optimal performance the correct coverslips should be used. One should also use the correct immersion oil for a given lens manufacturer as they are not all exactly the same refractive index and this can lead to spherical aberration.

As well as the NA and magnification of the lens another very practical consideration is the working distance. This is the distance from the end of the objective to the focal point. If the lens is designed to work through a coverslip (for example oil objectives) then the working distance is measured at the actual surface of the specimen (in other words the coverslip is already allowed for and not quoted in the distance). As a general rule the higher the NA and magnification the shorter the working distance. In the more expensive high NA lenses with very short working distances, the end of the lens is normally spring loaded to prevent damage to either the lens or sample.

The axial range through which the image is in sharp focus is known as the depth of field. This varies significantly depending on the lens numerical aperture and also to some extent with the magnification. The depth of field is given by

$$d_{\text{field}} = \frac{\lambda n}{\text{NA}^2} + \left(\frac{n}{M \times \text{NA}} \right) e$$

where the terms have the same definition used earlier, M is the lens magnification and e is the pixel size on the detector (in the case of a digital camera) or the smallest distance that can be resolved by the eye (in the case of direct observation). It should be noted that in the case of a confocal microscope this equation is no longer valid, as is explained in Chapter 5. These various properties of lenses are summarized in Table 3.2 and although the table contains only a selection of Nikon lenses all manufacturers have similar magnifications, NA and working distances as these are fundamentally set by the mechanical and optical

TABLE 3.1   Naming and Correction Features of Microscope Objectives

Lens Type	Cost	Chromatic Correction	Spherical Correction	Notes/Use
Achromat	●	Blue & red	Green	Green, monochrome, viewing
Fluorite	●●	Blue, red, part green	Blue, Green	Colour photography in white light
Apochromat	●●●	Deep blue, blue, green, red	Blue, green red	Colour viewing and fluorescence
Plan-fluorite, Plan-apochromat	●●●●	As fluorite, or apochromat	As fluorite, or apochromat	Flat field, recording (film or digital), scanning

TABLE 3.2 Summary of Typical Microscope Objectives (These Exact Lenses Are Manufactured by Nikon but All Manufacturers Produce Lenses with Very Similar Properties)

Lens Type	Magnification	Numerical Aperture	Working Distance (mm)	Maximum Lateral Resolution at 500 nm (μm)	Depth of Field (μm)
Achromat					
	×4	0.10	30	6.1	47.2
	×10	0.25	7.0	2.4	6.7
	×20	0.40	3.9	1.5	25
	×40	0.65	0.65	0.9	0.8
	×60	0.80	0.30	0.8	0.6
	×100 (oil)	1.25	0.23	0.3	0.1
Plan Achromat					
	×1	0.04	3.20	15.1	305.6
	×2	0.06	7.50	10.1	134.3
	×4	0.10	30.00	6.1	47.2
	×10	0.25	10.50	2.4	6.7
	×20	0.40	1.20	1.5	25
	×40	0.65	0.56	0.9	0.8
	×100	0.90	0.26	0.7	0.5
	×100 (oil)	1.25	0.20	0.3	0.1
Plan Fluorite					
	×4	0.13	17.20	4.7	26.8
	×10	0.30	16.00	2.0	4.3
	×20	0.50	2.10	1.2	1.3
	×40	0.75	0.66	0.8	1.3
	×40 (oil)	1.30	0.20	0.3	0.1
	×60	0.85	0.30	0.7	0.5
	×100	0.90	0.20	0.7	0.5
	×100 (oil)	1.30	0.16	0.3	0.1
Plan Apochromat					
	×2	0.10	8.50	4.0	29.4
	×4	0.20	15.70	2.0	5.8
	×10	0.45	4.00	0.8	0.1
	×20	0.75	1.00	0.5	0.1
	×40	0.95	0.14	0.4	0.1
	×40 (oil)	1.00	0.16	0.4	0.1
	×60	0.95	0.15	0.4	0.1
	×60 (oil)	1.40	0.21	0.3	0.1
	×60 (water immersion)	1.20	0.22	0.4	0.1
	×100 (Oil)	1.40	0.13	0.3	0.1

constraints of a microscope. The table does not include more specialized lenses which have variable apertures and working distances (which also affect the NA) nor ones suitable for multiple media.

The engraving and colours on an objective lens provide the vital information on its use. The first engraving will be the manufacturer's name, then the lens specification as the magnification and NA. There may also be a comment on its use for DIC or other more specialized imaging methods. This information is then followed by the lens back focal distance (∞ or 160), the coverslip thickness and the working distance. In addition many lenses have one or two colour bands. The thicker coloured band nearer the mounting thread indicates the lens magnification.

1–1.5×	Black
2–2.5×	Brown
4–5×	Red
10×	Yellow
16–20×	Green
25–32×	Turquoise
40–50×	Light Blue
60–63×	Bright Blue
100–250×	White

If present a second band displays the immersion medium for that lens, white for water, black for oil, orange for glycerin and red for a special or other immersion fluid. No band indicates air.

As a summary guide the following should be considered in order to help with the correct selection of the best lens for a specific task.

- Highest NA possible for best resolution

- Correct immersion fluid (water or oil) or air

- Select working distance

- With or without coverslip

- Magnification to obtain as large an image as required on the detector

It is always worth reading the microscope manufacturer's table of lenses as new lenses for specific applications are being developed continually. As will be discussed in the next chapter certain lenses are also optimized for more complex forms of microscopy, though they are also suitable for basic widefield imaging.

It is also worth noting here the best method of cleaning optics, in particular objective lenses. Oil objectives should be cleaned after use to ensure there is no risk of oil entering the objective. Cleaning should be undertaken using suitable lens cleaning tissue (available from most microscope suppliers and optics companies). Small particles of dust can be blown off the surface of optics (taking care of delicate elements) using a specialized can of compressed clean air (not from the machine shop or compressor). If using lens tissue this should ideally be moistened with a suitable solvent such as high purity isopropanol (IPA), acetone, methanol or ethanol. A single drop on a tissue is normally sufficient and this should be gently dragged across the optic in one direction and then discarded. If the optic is still dirty then use a second clean tissue in the same manner. The cost of lens tissue is small compared to the cost of the optics! Water dipping objectives should also be cleaned in a similar manner after use to minimize the risk of salt crystals appearing on the lens from the immersion solution.

3.6 IMAGE DETECTION AND RECORDING

Having illuminated and imaged the sample, the image now needs to be viewed. The light from the microscope objective may pass through some intermediate optics in the more advanced methods described in Chapter 4. However, in a widefield system the light passes through a lens and onto the detector. The most commonly used detecting system is still the observer's eyes. These highly sensitive detectors are capable of visualizing in colour, have a huge dynamic range and integrated signal processing! Their main limitation being that they cannot record a permanent record of the image. The value of the magnification and numerical aperture of the eyepieces in a microscope are again important. If one requires an overall magnification of say ×200 this could be achieved using either an objective of ×20 or ×10 magnification with ×10 or ×20 eyepieces. Typically, for high resolution, one would prefer a ×20 0.75 NA objective with ×10 eyepieces rather than a ×10 0.5 NA objective, ×20 eyepiece combination. The latter may give a wider field of view but with a lower overall resolution due to the lower NA of the lens.

In terms of permanent records in a widefield microscope one is generally left with either an electronic camera or the more traditional, and now basically obsolete, film-based system. Film actually provides the highest possible resolution with a wide dynamic range and a permanent record. Digital images are now the routine method used throughout the world for permanent records. The selection of the correct camera is, however, important to ensure that the maximum resolution of the microscope optics is actually recorded.

The lens before the camera is known as the tube lens and in order to have the correct magnification from the microscope objective this lens needs to be matched to the manufacturer's design specification. Table 3.3 lists the different focal lengths of tube lens expected

TABLE 3.3   Tube Lens, Thread Type and Parfocal Distance for the Main Manufacturers' Objective Lenses

Manufacturer	Objective Lens Thread Type	Tube Focal Length (mm)	Parfocal Length (mm)
Zeiss	RMS	165	45
Olympus	RMS	180	45
Leica	M25	200	45
Nikon	M25	200	60

Note: The parfocal distance is the distance from the back shoulder of the objective (the reference plane) to the focal point of the lens. For a single manufacturer these are constant so that when a lens is changed the microscope does not need to be refocused.

from each of the main manufacturers. The table also notes which thread the objective lens uses to attach to the microscope for each manufacturer as the use of the previously ubiquitous RMS thread (around 19 mm in diameter) was forsaken by some suppliers in the desire to improve microscope images when infinity optics were produced.

In selecting the camera the important considerations are the pixel size, the number of pixels, the speed of imaging required and the sensitivity and noise on the camera. There may also be a decision for colour or monochrome but as explained in Chapter 2, in many applications the preference is for a monochrome camera due to the presence of the Bayer filter in colour sensors. In order to obtain a diffraction limited image that is correctly sampled, one would like to have at least two pixels sampling the point spread function (psf) of the lens. This is to achieve the Nyquist imaging limit, which is discussed in Chapter 5. A ×20 0.75 NA lens has a point spread function of 0.4 µm, which means the spot will be 8 µm on the camera if one uses the normal tube lens for that objective. This would mean one would like pixels of 4 µm for Nyquist sampling. Indeed if one was going to undertake any image analysis to find the centre of features or track movement one might prefer the psf to be spread over perhaps nine pixels. In this case most camera pixels are not small enough and the option is to use a further lens, or alternative tube lens, with a focal length that magnifies the image onto the camera so that it covers the correct number of pixels. As the camera chip size is finite this will mean a loss in the field of view. In magnifying the image in this manner one also spreads the available light over a larger area reducing the signal at any one pixel. Higher speed cameras are frequently more noisy and thus should really only be used when their speed is required to record the dynamic motion within a sample. Camera selection involves trading-off these various factors for a specific imaging task.

Storage of images should also be considered at the outset of the selection process. Images should never be saved in a lossy compressed format such as JPEG or GIF. Significant information in an image is lost when this compression takes place (see Chapter 13). The preferred option is normally a TIF or BMP file. The number of images to be stored (in particular for dynamic applications) can lead to significant data handling and storage challenges and thus should be considered in any imaging protocol.

3.7 GUIDELINES FOR USE AND ADVANTAGES AND DISADVANTAGES OF A WIDEFIELD MICROSCOPE

Widefield imaging is the cornerstone of all optical microscopy. The basic instructions for use are as follows.

1. Select the most suitable lens for the imaging task. Generally this is going to be as high an NA as possible that will provide the required field of view, which is strongly determined by the magnification of the lens.

2. Adjust for Köhler illumination as described above.

3. Place the sample on the stage. At this point it may be useful to switch to a lower magnification lens to find the correct area of interest within the sample.

4. If using an oil objective place a small quantity of oil using a dropper onto the lens (for an inverted system) or the coverslip for an upright. Use as little oil as possible and be careful not to cause any bubbles.

5. Slowly adjust the focus to obtain the best image. Be aware that you may see dirt on the coverslip or slide and that this can help to adjust the focus up or down if one knows which plane one is currently focused upon.

6. Switch over to the objective for the final imaging if a lower magnification was being used to search through the sample. Although all systems should be parfocal (all objectives in focus on the same plane) check for fine adjustments to the focus.

7. If required switch over to the camera to record the image ensuring all images are saved in a lossless manner (not GIF or JPEG).

8. Make a note of the microscope parameters being used such as the lens, light source, camera settings. A text file or spreadsheet saved on the computer in the same directory as the images is a useful way of keeping track of this "meta-data" if it is not automatically recorded by the microscope.

9. As a basic guide to using a digital camera one should try to keep the gain as low as possible as this will help to reduce the noise. However, the illumination light should not be set at too high a level to minimize the risk of sample damage. The exposure on the camera can be increased though clearly this may be a limiting factor when one wishes to observe dynamic events. One can also "bin" pixels on some cameras where the signal from four, or more, pixels is combined before being digitized. This can help in situations where the noise on the camera is dominant. Pixel binning clearly reduces the resolution of the recorded image, but it can be a useful technique to employ when moving around the sample if the integration time of the camera is long as it will reduce the exposure time.

10. Save the images remembering not to use a lossy compression format.

11. At the end of the imaging session switch off the illumination, clean up the area and in particular remove any oil immersion objectives from the microscope and carefully remove the oil from the objective and return the lens to its container, storing the lens with the lens pointing down. This reduces the possibility of any oil slowly seeping into the elements of the lens which, although unlikely, can happen and will lead to expensive repairs, or the purchase of a new lens.

12. Copy any images and save a backup so data is not lost.

It is also worth having a microscope slide with a grating, or other features, etched into the surface for calibrating and checking the magnification of the microscope. Suitable slides are sold by most manufacturers or optics companies and consist of lines at a known spacing. One such target is known as the USAF test target, which has pairs of lines at different spacings. These can then be imaged and the resulting image analyzed to ensure that the spacing determined by the software is that known from the test target information sheet. Although this should not change over time once a system has been calibrated, whenever a new lens is used it is always worthwhile checking that the spatial calibration is being correctly recorded.

One can also use such a sample to check the resolution of the imaging system. Using a sample that has a sharp straight edge from light to dark in the image, one records a single image. This image can then be analyzed and a single intensity profile drawn through the edge from light to dark region. If the imaging system were perfect this would produce an instantaneous change in intensity. However, diffraction means that this change will appear as a slope, though it could be steep. If the line is then differentiated one should end up with a curve that will be Gaussian in shape. Measuring the full width, half maximum of the curve will provide the resolution of the system. This can be undertaken either using a home written program or the line profile can be imported into a spreadsheet with the pixels on the x-axis and intensity on the y-axis. The differentiation can then be undertaken by taking the difference between the intensity of each adjacent pair of pixels and the answer provides the slope (change in y divided by the change in x or dy/dx). This can then be plotted against the pixel value to produce the required curve.

Nearly all microscopes discussed in this book will have the option to operate in the widefield basic mode. This method should be used, wherever possible, to undertake a quick inspection of the sample before using more complex methods. Using such a system, in particular with a low magnification (×10) lens, at the start of inspecting a sample is very helpful to locate the areas of interest. Widefield imaging provides a large field of view and does provide information at different depths, even if some of that information is blurred due to the depth of field of the lens. In the life sciences the widefield system is excellent at looking at layers of single cells or thin sections of fixed tissue. Reflection microscopy in the widefield configuration is excellent at looking at surfaces.

The disadvantages are the potentially poor image quality in thick samples where the out of focus transmitted light can significantly lower the contrast in the sample. It can also be difficult sometimes to see single cells, which are basically transparent. One requires a contrast feature and this is where some of the more advanced methods described in Chapter 4 start to be applicable.

In summary, widefield microscopy is excellent for imaging a large area at high speed with good resolution in the optical plane. Dynamic changes can be observed and recorded with ease, as long as they are in one plane. One does, however, require some attribute of the sample to provide a level of contrast, though the same can be said for all optical microscopy methods.

CHAPTER 4

Advanced Widefield Microscopy

4.1 INTRODUCTION

In Chapter 3 the basics of widefield microscopy were introduced with a focus on the optical components and systems used. It was assumed that the sample contained sufficient contrast in reflection or transmission to observe the features of interest. However, in many imaging challenges this is not the situation and more advanced methods are required. In this chapter a wide range of such advanced techniques are described demonstrating how they convert optical properties within the sample into contrast in an image. All the methods produce widefield images, which can be observed using the operator's eyes and then stored on a camera. Generally all methods are capable of high speed imaging, though none provide any optical sectioning.

The techniques described in the chapter are available as minor additions to the basic widefield systems described in Chapter 3. They are all available as enhancements to commercial systems and can be added either at the time of purchase of the main microscope, or as extra features at a later date. Some of the methods require special objective lenses but as these will all operate as conventional widefield lenses, they can be considered during any original purchase though they may be slightly more expensive than a base lens with the same magnification and NA. Only one of the methods in this section, fluorescence, requires any addition or alteration of the sample and in some cases fluorescent compounds may already be present.

4.2 POLARIZATION MICROSCOPY

The concept of polarization was introduced in Chapter 2 Section 2.1.4 where it was stated that light waves that are linearly polarized have their electric fields oscillating in the same direction (Figure 2.2). As light passes through most materials the polarization of the light field is not altered. However, under certain conditions and through certain materials this is not the case and this is when polarization microscopy is used and can provide unique information on the sample.

Scattering of light is one phenomenon that can lead to a change in the polarization of light. Direct reflection from a surface does not change the polarization but, as the scattering process is a result of light interacting with a molecule's or particle's electric field, when the light is re-radiated it can have its polarization altered depending on the alignment of the molecular field. This effect is further discussed in Chapter 10. In reflection microscopy polarization can thus be used to separate out the direct reflection from a surface from the light backscattered from deeper within the sample. In transparent samples, which contain scattering particles, polarization can be used to separate out the particles from the surrounding material. One area where this has been used is in optical tweezers to enhance the contrast between the suspended particle and the surrounding fluid.

The other effect that can alter the polarization of light in an optical microscope is birefringence. This occurs in certain materials (generally crystals and plastics) but can also be present in suspensions of particles. A material that is birefringent has a refractive index that depends on the polarization and propagation direction of the light. This is best understood by considering a crystal. Along a certain optical axis of the crystal light that is of one polarization will travel faster than that of the opposite polarization. Light polarized in exactly this plane will travel through with a single speed and the birefringence will not be noticed. However, if we have light polarized at an angle to the birefringent axis we have to consider the light as being made up of two "components", one in line with the birefringent axis and one which is not. The light that is in line will experience one refractive index and travel at a higher speed than the component which is along the slower optical axis. This means that when the light re-emerges from the sample part of the wave will have been delayed relative to another and this leads to both constructive and destructive interference leading to light and dark areas within the emerging light field which can be seen by eye or on a camera. For light with multiple wavelengths some colours will have been delayed more than others leading to colours appearing in the image.

4.2.1 Practical Implementation of Polarization Microscopy

In order to undertake polarization microscopy only minor additions are required to a conventional widefield microscope. Figure 4.1 illustrates the practical components in a polarization microscope. The light from the illumination system is passed through a polarizing component, which only permits one polarization of light through the system. This component is normally achromatic to ensure that a white light source can be used in the microscope though a narrow spectral filter and associated polarizer can also be used. The polarizer is normally aligned so that the transmitted light wave is oscillating in the east–west direction across the sample.

The light then continues through the sample and objective before it encounters a second polarizing element, sometimes known as the analyzer. This can be adjusted to be crossed with the previous polarizer so that only light that has had its polarization altered passes through to reach the detector. For obvious reasons the optics within the system should not add any local variations to the light and thus it is important to use strain-free objectives and lenses that are marked as being suitable for polarization microscopy.

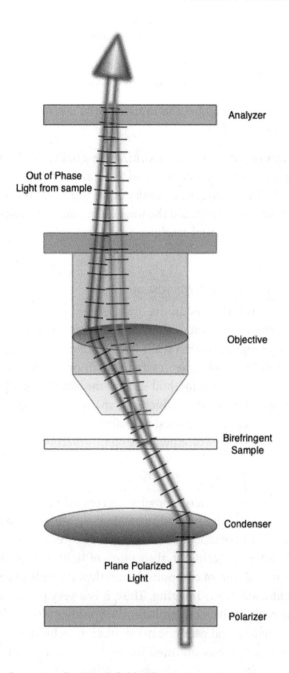

FIGURE 4.1   Basic configuration for a widefield polarization microscope.

The following provides a simple set-up procedure for polarization microscopy.

1. Set up the widefield microscope for Köhler illumination.

2. For a transmission system place a suitable microscope slide on the stage, or place the sample on the stage but position it so that the light is passing through a clear area of the slide.

3. For a reflection system place a mirror in the sample position.

4. Adjust the analyzer so that no light is seen on the detector or through the eyepieces. This ensures that the polarizers are correctly crossed.

5. Remove the test area and adjust the sample to the correct position on the stage and observe the image.

6. The analyzer can be adjusted to maximize the quality of the features seen in the resulting image but should not be adjusted to obtain a brighter image. If the light level is too low consider increasing the intensity of the illumination or the integration time on the camera being used to record the image. One may also consider "pixel binning" which will reduce the potential resolution of the recorded image but can increase the recorded signal to noise.

4.2.2 Applications of Polarization Microscopy

Polarization microscopy is little used in the life sciences because the samples are not normally significantly birefringent because they contain large quantities of water. The method is used extensively in geology and earth sciences and other fields where crystals are present in the sample. It is possible to observe crystal growth or melting using the polarization method and indeed it is a standard method used to measure the melting point of crystals. Certain polymers are also birefringent which can change when the sample is under stress. Polarization microscopy can then be used to examine the local strain in detail, and it is possible to make quantitative measurements using this method.

4.3 PHASE CONTRAST MICROSCOPY

As a technique, phase contrast microscopy has decreased in popularity with the advent of fluorescent labelling methods, three-dimensional imaging and advanced computing. However, it remains an important method by which to examine basically transparent samples, which have little scattering or absorption of light to provide inherent contrast. It does not require any addition of compounds and thus the only external perturbation to the sample is the light used in the imaging. Thus, it is a very powerful tool for looking at live samples in their native state, though lacks the specificity that can be achieved using modern fluorescent labelling and genomic manipulation techniques.

Phase contrast microscopy was invented in the 1930s by Frits Zernike, a Dutch theoretical physicist specializing in optics. He was examining the way light travelled through objects and in particular how the wavefront was distorted due to local changes in the refractive index. This work led to his development of Zernike polynomials, which were discussed in Chapter 2 in relation to optical aberrations. Zernike then considered a practical optical method by which the changes in phase of the wavefront, caused by refractive index changes, could be visualized. Cameras and the human eye cannot directly detect the phase of light, only its intensity. For this work Zernike was awarded the 1953 Nobel Prize for Physics.

As discussed in Chapter 2, when light travels through any material it is slowed down compared to light that travels through a vacuum. The ratio of the speed of light in the material to that in a vacuum is known as the refractive index of the material. What Zernike realized was that if a plane wave of light could be split into two, which travelled through areas of different refractive indices, the waves would become out of phase. If these beams could then be recombined they would interfere either constructively or destructively depending on the exact optical distance difference through which they travelled. The optical distance is the physical distance travelled by the light multiplied by the refractive index. Thus, the contrast in the images will be produced by differences in the optical path and in thin samples this will be directly related to the local refractive index. His invention was to incorporate a very simple way of achieving this effect into a microscope.

4.3.1 Practical Implementation of Phase Contrast Microscopy

The key concept in phase contrast microscopy is to separate out light that passes through the area around the sample and that, which passes through the area of interest. In the sample, where there are differences in the local refractive index, the light is diffracted. The light from the two different paths is then recombined on the detector. Figure 4.2a illustrates the main features for the method as implemented in a standard microscope. The light that passes around the sample is normally attenuated so that it does not dominate the final image, and a known phase advancement, or delay, of $\lambda/4$ is added to this light to further help improve the final image contrast.. To achieve this in an optical microscope two extra components are added to the basic optical system described previously. The first is an optical annulus placed such that it is re-imaged onto the back aperture of the objective (aperture conjugate planes), while the second, a phase plate, is placed at the back aperture of the objective lens. In undertaking phase contrast microscopy the light source must be well adjusted for Köhler illumination to ensure high contrast images.

In Figure 4.2a the light, from the collection lens in front of the bulb in a conventional Köhler configuration, passes through the first aperture in the illumination system and then encounters the new optical annulus. The light subsequently reaching the sample is sometimes described as a hollow cone of light. However, this is not strictly speaking accurate although it does provide an easier picture to understand. Light from this cone that passes through the sample undeviated subsequently arrives at the rear focal plane of the objective as a ring of light. Light that is diffracted by the sample, which is much fainter, is spread over the whole rear focal plane of the objective. If these two parts of the illumination were then allowed to recombine at the detector the diffracted light would be around $\lambda/4$ behind the direct light. Any interference pattern present would be masked by the significantly higher intensity of the undeviated light and in addition the interference would not be totally destructive as the delay is only around $\lambda/4$. Thus to increase the contrast the undeviated light passes through a thinner material on the phase plate such that it will arrive $\lambda/2$ ahead. This light is also attenuated so the intensity is roughly that of the deviated light. When the two sets of light rays are now combined they will produce an interference pattern which can clearly be seen and where the pattern is based upon the optical thickness of the sample. It is also possible to delay the undeviated light by $\lambda/4$ in which case the interference pattern is constructive, leading to a bright image, so called "negative phase contrast".

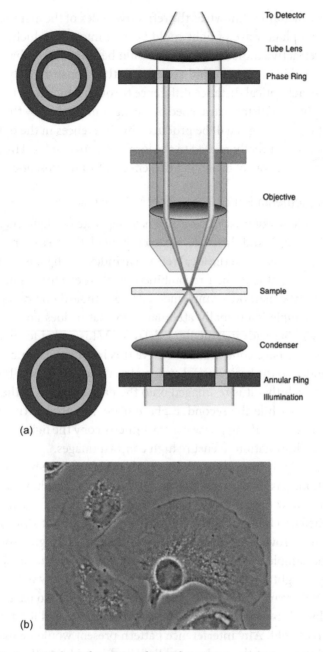

FIGURE 4.2   (a) Basic optical configuration for a phase contrast microscope, (b) a typical phase contrast image (from data sheet by permission of Nikon Instruments).

Clearly the phase delay needs to be optimized for one wavelength and this is at 546 nm, which is the green light produced by a mercury lamp. If chromatic light is used, i.e. without a green filter, then the images detected will have additional colour features present due to the different retardance for the range of wavelengths reaching the detector. This can enhance features within an image though generally the overall contrast of the image is lower.

Typically the annular aperture is provided in microscopes as a standard grid in the illumination system. The phase plate and dark ring is normally integrated into special phase objectives. These objectives can be used for normal imaging with only a minor decrease in the image quality but the preference is to use them only for phase imaging.

The following provides a simple set-up procedure for phase contrast microscopy.

1. Set up the widefield microscope for conventional Köhler illumination.

2. Place a sample on the microscope (cheek cells provide a readily available test material!) and focus on the sample.

3. Open the condenser aperture diaphragm and insert the green filter into the illumination path along with the correct aperture mask.

4. Remove one eyepiece and insert the "phase telescope" or "Bertrand lens" that will be supplied with the microscope (this will focus on the back aperture of the objective).

5. Using the phase telescope observe the back aperture of the objective and with the annulus adjustment screws centre the annulus in the field of view onto the phase mask integrated into the objective. It can sometimes be worth removing the sample at this point.

6. Once the annulus and phase mask have been aligned remove the phase telescope and return the eyepiece and sample if this has also been removed. Adjust the focus on the sample.

7. This process has to be repeated for each objective lens.

4.3.2 Applications of Phase Microscopy

Phase contrast is an excellent method to observe unlabelled, thin and nearly transparent samples such as individual cells or tissue cultures. The method provides the ability to image rapidly and monitor changes in cell thickness or movement as both change the optical path and hence alter the observed light pattern. Phase contrast is also insensitive to polarization and birefringence effects, which is a major advantage when examining living cells growing in plastic tissue culture vessels.

There are, however, a few limitations in phase contrast microscopy:

a. The numerical aperture of the lens (and hence ultimate resolution) is reduced slightly as the phase mask reduces the back aperture and hence the effective numerical aperture of the lens.

b. Phase images are frequently surrounded by "halos" of light, which are artefacts of the imaging process and can distort boundary details.

c. Phase contrast is not suitable for thicker samples due to phase shifts that occur naturally above and below the focal plane.

d. Images, if using white light, often appear grey and slightly lacking in contrast. This is now less of a problem using CCD cameras compared to previous use of film, as the images can be post processed. Colour artifacts can also obscure features if using white light. Green only illumination can also lead to some distortion of the image on colour cameras due to the Bayer filter used.

4.4 DIFFERENTIAL INTERFERENCE CONTRAST (DIC) MICROSCOPY

Differential interference contrast (DIC) microscopy has some similarities to phase contrast microscopy in that it relies on changes in the optical path length for contrast in the final image. However, in DIC the contrast is provided by the gradient of the optical path length difference whereas it is the absolute optical path length difference that appears in phase contrast. They are therefore complementary techniques and the advantages and role of each method is discussed at the end of this section.

DIC microscopy was demonstrated in 1955 by Francis Smith. His method used a Wollaston prism and polarization of light to produce two light beams to pass through the sample. These were subsequently recombined to produce an interference pattern. The gradients in optical path length difference then appeared as light and dark features in the transmission image. This core concept was later improved as a practical microscopy method by George Normarski, who invented an enhanced Wollaston prism (Nomarski prism) that is thin and is less critical to the position it has to occupy in the optical train within a microscope.

The optical concept of a DIC microscope is shown in Figure 4.3a and consists of two Wollaston prisms between crossed polarizers. In a Wollaston (or Nomarski) prism incoming light is separated into two linearly polarized output beams, which are physically displaced from each other (by less than the diffraction limit of the microscope) and have orthogonal (opposite) polarizations. Although the two beams are travelling through the sample very close to each other they do not interfere as they have different polarizations. However, each beam takes a slightly different path through the sample, and thus a phase difference between the two beams will develop. The greater the difference in optical path length between the two closely spaced rays, the greater the contrast in the image. After passing through the sample the beams are sent through a second prism which brings the polarizations back together again. They are now able to interfere leading to contrast in the image. As the two rays are physically close to each other through the sample they are effectively looking at the gradient of the path difference rather than an absolute path difference between the paths. In order to ensure that it is light that has interacted with the sample the pair of Wollaston prisms are mounted between crossed polarizers. Only light that has experienced a slight polarization change as it passes through the optical system and sample will then be visible.

Light from the illumination source first passes through a linear polarizer, which is typically aligned in an east–west direction. The resulting beam then passes into a modified Wollaston prism (Nomarski prism) from which the two perpendicularly polarized beams emerge,

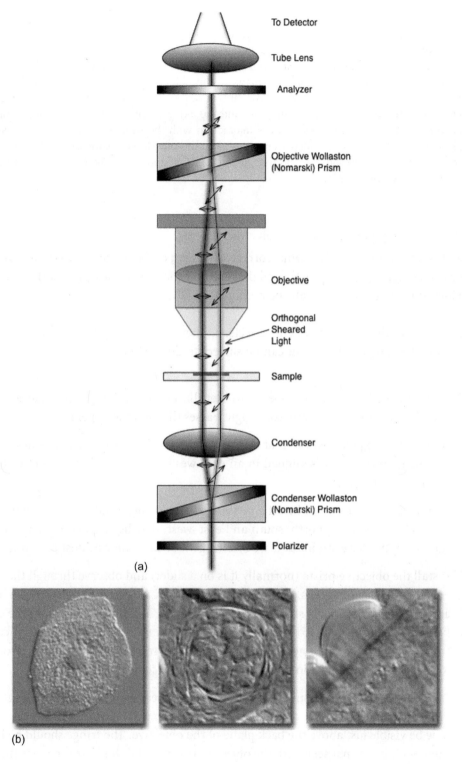

FIGURE 4.3 (a) Basic optical configuration for a DIC microscope, (b) typical DIC images (from Microscopy U, by permission of Nikon Instruments).

subsequently intersecting at the front focal plane of the condenser. They then pass through the sample parallel to each other and suffer slightly different phase delays due to their individual optical paths through the sample. When these parallel beams enter the objective lens they are focused above the rear focal plane of the objective where they enter the second Wollaston prism. This combines the two beams at a specific distance outside the prism, and removes the optical path length difference introduced after the first prism; any path length difference remaining will be due to the interaction with the sample. The light then passes through the second analyzer, aligned north–south, and onto the detector. The system is achromatic, except for the sample, and thus any colour that appears in the image is due to differences within the sample.

4.4.1 Practical Implementation of DIC Microscopy

As with all precision optical systems, correct alignment of the microscope is crucial to high quality imaging. After ensuring that all the required components are present the following provides a starting point to obtaining high quality DIC images.

1. Ensure all the DIC components are clean. They are normally kept inside the microscope housing on sliders but can be subject to dust. Mount all the components into the system.

2. Set up the widefield microscope for Köhler illumination, though the polarizers may need to be adjusted to ensure some light passes through the system.

3. Adjust the two polarizers so that they are orthogonal to each other. By tradition the illumination polarizer is aligned in an east–west direction. When correctly aligned the field will be dark.

4. Remove one eyepiece and insert a phase telescope or Bertrand lens and a dark cross should be seen facing north–south and east–west. Any light spots that appear indicate the optics have strains and are birefringent or there may be dust present.

5. Install the objective prism (normally it is on a slider) and observe through the phase telescope or Bertrand lens. The back aperture of the objective should now be bright and uniform except for a single dark interference fringe extending across the entire field of view at around 45° along the sheer axis. In some systems the prism can be translated using a small adjusting knob and this moves the fringe away from, or towards, the central position. Position the fringe as centrally as possible.

6. Remove the objective prism and insert the condenser prism. Some refocusing of the phase telescope or Bertrand lens may then be required but the fringe pattern should now be visible just above the back plane of the objective. The fringe should be in the same position as that seen with the objective prism and if this is not the case the condenser prism should be adjusted to achieve this effect.

7. Remove the phase telescope or Bertrand lens and replace the eyepiece.

8. Replace the objective prism and place a thin transparent sample onto the microscope (a cheek cell is a suitable sample). While observing through the eyepieces make minor adjustments to the analyzer to obtain the best dark field. Then focus carefully on the sample.

9. Adjust the upper prism (known as the bias retardation) to set the contrast for the best quality image for the specific task being undertaken.

An alternative to the Wollaston or Nomarski prism is known as the de Sénarmont configuration; the method is similar but the adjustment of the prism is slightly different and the specific manufacturer's instructions should be followed.

4.4.2 Applications of DIC Microscopy

DIC imaging is excellent for looking at single layers of cells but it should always be remembered that the image, although it may have the appearance of a surface image, is the integrated path difference throughout the cell. Samples which have sharp changes in optical thickness are excellent for DIC as the phase gradient is large. Due to the nature of the polarized light in a DIC system samples can sometimes be rotated around the direction of the polarization of the light to enhance certain features, for example the sarcomeres within muscle cells.

DIC has several advantages over phase microscopy. One of these is that one uses the full back aperture of the objective and hence obtains better resolution due to the higher numerical aperture. The method also enables some sense of the three-dimensional nature of a sample to be explored, partly due to the higher NA but also due to the shadowing that appears in DIC images. There are also none of the halo effects seen with phase methods. The images can produce significant colour variations enhancing certain features. Generally all plan-achromat and achromat lenses can be used for DIC unless otherwise stated by the manufacturer.

However, there are some disadvantages. The components are more expensive and care has to be taken in selecting the correct tissue culture and observation chambers as the images are affected by any birefringent properties of the materials present in the optical path. Alignment is also significantly more critical and some apochromatic objectives are not suitable as they can affect the polarization of the light. Phase microscopy is generally better for flat samples with fewer sharp variations in the optical depth.

4.5 DARKFIELD MICROSCOPY

In all the methods described so far the aim has been to illuminate the sample with light and to maximize the collection of the light by the objective. In darkfield microscopy the central part of the illumination is obscured and only the peripheral rays enter the sample. These will generally miss the objective lens, but if the sample has features which will scatter, refract or reflect, some of the photons will be directed into the objective and hence will be seen. This results in bright features appearing on a dark background and this can frequently enhance specific structures within the sample.

4.5.1 Practical Implementation of Darkfield Microscopy

The basic configuration is shown in Figure 4.4a. The system is based upon a conventional illumination configuration but a light stop is placed in the centre of the illumination path. This means the illumination optics produce a hollow cone of light. This light cone passes through the sample and most will misses the collection objective. Thus only light deviated by features in the sample will enter the imaging system.

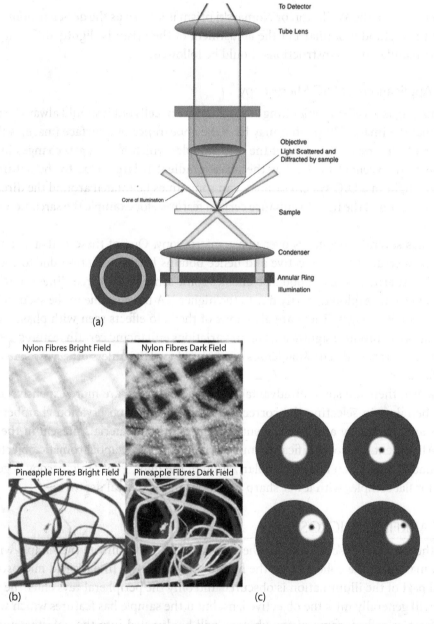

FIGURE 4.4 (a) Basic configuration for darkfield microscopy, (b) typical transmission and dark-field images (from data sheet by permission of Nikon Instruments), (c) image seen when aligning a high NA darkfield condenser.

If the darkfield optical system is considered in terms of Fourier optics the light stop removes the zeroth order (unscattered or undeviated light) from the diffraction pattern that is formed at the rear focal plane of the objective. The peripheral, or oblique rays, which do interact with the sample will be diffracted and the first, or higher, orders of diffraction will be collected by the objective lens and are then re-imaged onto the detector by the remaining optics in the system. The image is thus made up of only the highest order features in the diffraction pattern.

The resolving power of the system has not changed but the images can appear to be very different. This is because objects which have smooth surfaces, or small changes in refractive index, only produce small angular changes in the light path and thus few of these rays will be collected by the objective. Sharp features, which cause significant levels of scattering, however, have a much higher chance of passing into the lens and subsequently appearing in the final image. The method is again suitable for unlabelled samples, which may have sharp features as illustrated in Figure 4.4b.

The alignment in a darkfield system, as with all optics, can significantly affect the performance and the following procedure provides a basic guide for good image quality. As before, individual systems can require marginally different operating procedures. The following is designed for low magnification darkfield transmission imaging.

1. Set up the widefield microscope for Köhler illumination.

2. Open both the field and condenser apertures to their maximum.

3. Select the correct stop for the objective in use (typically around 16 mm for a ×10 objective). Place this into the correct holder (often known as a spider holder due to the thin "legs" that support the aperture in the light path).

4. Insert the holder into the correct position in the illumination path.

5. For certain condenser system an additional low magnification lens may be used.

6. Remove the eyepiece and insert a phase telescope or Bertrand lens in place of the eyepiece.

7. One should see a bright disc of light that is partially blocked by the central stop at the back focal plane of the objective. The central stop diameter should be slightly larger than the bright disk and central in the field of view. If not, adjust the position of the opaque disk using the centring screws on the condenser.

8. Replace the eyepiece and place the required sample on the stage. One should have a bright image on a dark background. If the image is too dark increase the illumination intensity.

9. If required, the focusing condenser can then be adjusted to maximize the contrast in the image.

For high magnification, high NA objective lenses the adjustment is slightly different as a special high NA condenser lens is used. The following provides a route to correct alignment.

1. Using a low NA objective lens (×10, 0.25 NA for example) set up the widefield microscope for Köhler illumination.

2. Lower the condenser lens and replace the lens with a reflective high NA condenser. Raise the condenser to within around 2 mm of the sample stage. Some condenser lenses may require oil and apply in the normal way.

3. Place a suitable darkfield sample onto the microscope stage and an image similar to those in Figure 4.4c should be seen when using the ×10 0.25 objective.

4. As the condenser lens is focused image (a) should be seen. As the focus of the condenser is adjusted images similar to (b) should be observed. If image (c) is seen then the centring of the condenser should be adjusted. Image (d) indicates incorrect Köhler alignment and step 1 should be repeated.

5. If the condenser is an oil lens then the eyepiece can be removed and a phase telescope used in the eyepiece to ensure no air bubbles are present in the illumination oil as these can cause artefacts in the image.

6. The high NA observation objective can now be used and darkfield images obtained. Minor adjustment of the condenser lens may further improve the image quality.

4.5.2 Practical Applications of Darkfield Microscopy

Darkfield microscopy is sometimes used for imaging moving bacteria in water as their small size can cause them to create "sparkles" of light. The method is also useful for other samples with sharp features that are otherwise transparent.

4.6 FLUORESCENCE MICROSCOPY

Fluorescence microscopy is probably the most widely used microscopy technique in life science research and is popular in many other fields. The ability to specifically label features within the sample, most recently using genomic-based methods, provides an unprecedented advance in precision. Many of the methods described in later chapters are based upon fluorescence imaging. If a sample is going to be imaged using confocal or multiphoton microscopy it is always worth checking the sample first using conventional widefield fluorescence microscopy. The basics for fluorescence excitation are presented in Chapter 2 Section 2.4.4. The core of fluorescence imaging is therefore to deliver high intensity excitation light to the sample at one wavelength, and then to detect the resulting longer wavelength light (fluorescence) while rejecting the excitation light.

4.6.1 Practical Implementation of Fluorescence Microscopy

Figure 4.5a illustrates the basic configuration in an epi-fluorescence optical microscope. "Epi" means that the excitation light is delivered through the imaging objective and the

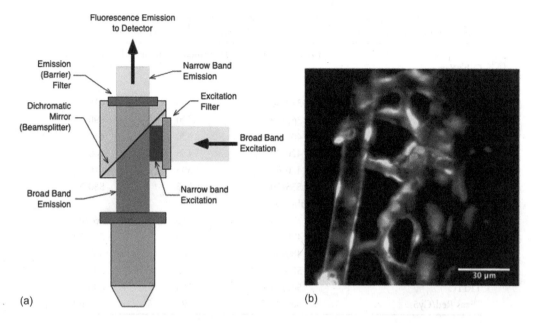

FIGURE 4.5 (a) Basic configuration of fluorescence microscopy, (b) multi-labelled fluorescence microscopy, GFP expressing vasculature and red expressing renin producing cells in a zebra fish (image courtesy of Dr Charlotte Buckley, Edinburgh University).

resulting emission light is collected by the same objective. Although transmission fluorescence microscopy is undertaken, or the sample is illuminated from the side, the epi configuration is by far the most common one used. Light from a suitable source (see Chapter 2 Section 2.2.5) is initially collimated and then passed through a "cut-off spectral filter". This has a sharp transmission band, which only passes shorter wavelengths of light, ensuring that there is no light hitting the sample which might be at the emission wavelength of the fluorophores in use. Even for a source such as an LED, which is typically considered to have a narrow spectral band, a cut-off filter should be used as LEDs have a "tail" of emission that can extend several tens of nanometers to longer wavelength and into the fluorescent emission band. The light hitting this cut-off filter should be collimated as the spectral characteristics of such filters are affected by the angle at which the light hits the glass and coating.

After being spectrally narrowed, the light is directed towards the sample using a dichromatic mirror. This reflects the short wavelength excitation light through the objective and into the sample. The returned light then passes back up through the objective lens and is then transmitted by the dichromatic filter. The light may then pass through a second filter to ensure that no excitation light reaches the detector, as the excitation intensity is several orders of magnitude brighter than the emitted light. As with a conventional widefield microscope the image light is then focused onto a camera in the normal manner to produce the fluorescent image.

The spectral separating component ("filter cube") is normally mounted on a rotating carousel so that a sample can be labelled with multiple fluorophores and each excited in turn. Standard systems are normally mounted with the filter sets as shown in Table 4.1,

TABLE 4.1   Standard Filter Combinations for Life Science Fluorescence Microscopy (The Numbers Indicate the Centre Wavelength and the Full Width Half Maximum of the Filter)

Cube Name	Excitation Filter /nm	Dichromatic Mirror /nm	Barrier Filter /nm
General Long Pass Filters			
UV-2A	333/50	400	410
UV-1A	365/10	400	390
V-2A	400/40	430	440
BV-2A	420/40	455	460
B-2A	470/40	505	510
G-2A	535/50	575	580
Band Pass Filters for Specific Fluorophores			
DAPI	375/30	415	460/60
GFP-B	470/40	500	535/50
FITC	480/30	505	535/45
CY3	525/30	555	590/50
TRITC	540/25	565	605/55
Texas Red/Cy5	560/40	595	630/60

though the exact wavelengths and nomenclature may vary slightly. The first group of filters are generic, covering a wide range of excitation, and more specifically emission, wavelengths. The wavelength given for the dichromatic mirror is the point at which the transmission passes 50% and the barrier filter blocks all wavelengths shorter than this figure, transmitting only the longer wavelengths.

The second group of filters in Table 4.1 are for specific fluorophores. The major difference being that the emission filter has a narrow pass band rather than just transmitting all longer wavelengths. This means that if one were using a sample labelled with GFP and CY3, although the blue light used for exciting the GFP might excite CY3, albeit weakly, the barrier filter would block the CY3 emission but pass the GFP signal. The table only provides a guide to the filter combinations available, and microscopy companies, or those specializing in filters should be contacted directly for further information.

For the best possible fluorescent imaging the microscope objective should be selected with care. This is not only to select the optimal NA and magnification but also to ensure that it is spectrally corrected at the right wavelengths. The objectives with "fluor" in their name are specifically optimized for fluorescent imaging and should be selected for the wavelengths being used according the Table 3.1. In particular the selection is likely to be an apochromat or plan-fluorapochromat.

CCD or CMOS cameras discussed in Chapter 2 Section 2.6.2 are ideal recording devices for fluorescence microscopy. Compared to conventional widefield imaging, rather than potentially splitting light between the eyepieces and camera all of the light should be sent to the camera. The fluorescence intensity from a sample is significantly lower than the signal that might be seen in other forms of widefield imaging. As in all forms of microscopy consideration should be given to the selection of the camera based upon the exact application. Higher speed, and thus very sensitive cameras, are required for imaging calcium dynamics either using fluorescent probes or through the more recent GCamp genetically encoded calcium indicators. Such observations were normally made using electronic multiplied CCD cameras, but in the last few years these have become superseded by sCMOS cameras, which generally have lower noise levels. It is also worth noting that most fluorescent images are recorded with monochrome cameras as they are more sensitive. Colour can be added in post processing through the addition of multiple images recorded using different filters.

A general protocol for using fluorescence microscopy is given below.

1. Switch on the fluorescent excitation lamp. In the case of mercury discharge lamps the lamp should be switched on and allowed to warm up for around fifteen minutes to ensure a stable output (or according to a specific manufacturer's instructions).

2. Set up the microscope for conventional widefield imaging. In order to minimize the risk of photobleaching it is prudent to use conventional illumination to identify the areas of interest in the sample. Switch over to the fluorescent illumination and ensure that no extra optics (such as Nomarski prisms or polarizers) are in the optical path. Switch off, or block, the conventional widefield illumination.

3. Ensure that the fluorescence illumination is correctly aligned by placing a fluorescent sea under the objective. This may be a fluorescent plastic slide or a solution containing a fluorophore (Nile red works well due to its huge spectral coverage both for absorption and emission). The fluorescent sea should appear to be uniform in emission and if not check the alignment of the illumination source into the microscope according to the manufacturer's instructions. The uniformity can be tested using the camera and a suitable colour look-up table, which codes intensity as colour in the image.

4. If a fluorescent sea has been used now switch on the conventional illumination and use this to find the required place for the sample of interest. With the intensity of the fluorescence excitation minimized, switch over to the epi-illumination, switching off or blocking the conventional illumination.

5. Slowly increase the illumination until one has an image of suitable intensity. The illumination level should be kept as low as possible to minimize the risk of photobleaching, which can occur even in a widefield fluorescence microscope.

6. At the end of the imaging session switch off the fluorescent illumination unless another user is about to take over the microscope. With the exception of LED illumination fluorescent excitation bulbs are generally expensive and have a limited lifetime. Many lamps contain a clock, which indicates the length of time the bulb has been used, and hence when it might require replacing. In the case of mercury discharge lamps if another user is arriving within an hour it may be kept on as switching on and off the lamp also lowers its life.

4.6.2 Practical Applications of Fluorescence Microscopy

Fluorescence microscopy is used in many fields ranging from geology through to the life sciences, though it is in the latter area that it is used the most. In the case of geology the samples are frequently inherently fluorescent, though they can be subject to bleaching in the same way as fluorescent molecular probes, so again care should be taken in the illumination levels. In the life sciences the applications are very broad but fluorescence generally provides the options of labelling a specific feature within a sample. Most recently this has been through the use of genetically encoded molecules but it should always be remembered that when imaging one is seeing where the fluorescent molecule is now, not always where it is being produced. As has been stated several times already the most important aspect in fluorescent microscopy is to keep the illumination level as low as possible to minimize damage to the sample and also keeping any photobleaching of the fluorophore low.

4.7 SUMMARY AND METHOD SELECTION

All the methods described in this chapter are aimed at specifically increasing the contrast in a sample, and in the case of fluorescence to highlight or label specific features. As with all forms of microscopy one should consider the perturbation being caused to the sample. For example, although fluorescence may be an excellent way of seeing a specific feature in a cell one has added a fluorophore that is not normally present within the sample and perhaps the same processes could be observed using DIC or phase microscopy. The notes below provide some guidelines on where each method may be suitable.

Polarization: Enhancing stress visibility of strain within a sample (stress dependent birefringence), crystal structure and changes in crystals, samples with asymmetric polymers present. Minimal perturbation to the sample with no additional compounds needing to be added.

Phase: Thin transparent samples, areas with slow changes in the local refractive index. Differences in thickness with the same refractive index will be resolved. Challenge of halos appearing around fine features. Minimal perturbation to the sample with no additional compounds needing to be added.

DIC: This is generally preferred over phase contrast for the reasons described earlier though the contrast is provided by gradients in the optical path length not the actual optical path length. High-speed changes can be seen, as generally the images can be adjusted to be bright with high contrast. Minimal perturbation to the sample with no additional compounds needing to be added.

Darkfield: Transparent samples with highly scattering points or small structures which diffract the light towards the field of view. Minimal perturbation to the sample with no additional compounds needing to be added.

Fluorescence: Specifically labelling features within the sample and also the ability to watch dynamic changes. Chemically specific labelling is possible and even the difference between bound and unbound Ca^{2+}.

CHAPTER 5

Confocal Microscopy

T HE DEVELOPMENT OF THE practical beam scanned confocal microscope in the 1980s marks the start of the recent explosion in optical microscopy methods. In providing the researcher with the ability to image intact samples in three-dimensions, linked with the growth of local computer processing, an array of new approaches could be taken to experimental design. For example, samples did not need to be so thinly sectioned leading to a revolution in many scientific disciplines, but the change has been largest in life sciences. For the first time intact living structures could be imaged for extended periods of time and thus the true nature of biology could be explored as being a dynamic three-dimensional (3D) system. The ability to image with high resolution inside such systems led to the desire for a new range of labelling methods, targeted to specific biological processes. This new direction for the life sciences culminated in the award of the 2008 Nobel Prize for Chemistry (see Shimomura et al. 2008) for the development of genetically encoded fluorophores. Many of the most advanced optical microscopy methods that are covered in subsequent chapters can be traced back to the core optical configuration of the confocal microscope. A brief historical background to this development has been presented in Chapter 1. As is a common theme in many of the currently standard microscopy methods, the original invention (Minsky 1961) had to wait for the development of technology, including lasers, detectors and computers, before widespread use was possible.

This chapter starts with a look at the basic principles behind confocal microscopy and how the ability to develop optically sectioned images is possible. It then concentrates on the practical details of such systems including the optics of beam scanning microscopes and the selection of detectors for different wavelengths. Some guidelines are then suggested for the best way to prepare samples for use in a confocal microscope along with the best way to optimize the settings on the system in order to obtain the results required for specific experiments. A focus here is to minimize the perturbation to the sample such that one is looking at a system in a state as close to natural as possible. The chapter ends with some comments on the reconstruction and analysis of 3D data sets; however, further details on the imaging processing methods that might be used are expanded upon in Chapter 14.

5.1 PRINCIPLES OF CONFOCAL MICROSCOPY

Fundamentally a confocal microscope works by preventing light that comes from outside the focal plane of the objective lens from reaching the detector. This enables a complete "optical slice" of the sample to be recorded. The focus can then be adjusted to take a series of optical sections throughout the sample and these slices can then be reconstructed into a 3D image. Modern confocal microscopes are still closely based on the principles first proposed and demonstrated by Marvin Minsky. His insight was to introduce a pinhole into an optical system, which was in the same focal plane as the sample. The pinhole is thus said to be *confocal* with the focal plane of the objective. This is illustrated in Figure 5.1.

In this simplified diagram light enters from the left-hand side (solid line), is reflected off the beam splitter (in a typical biological confocal system this is a dichromatic beam splitter transmitting one wavelength and reflecting another), and is subsequently focused by

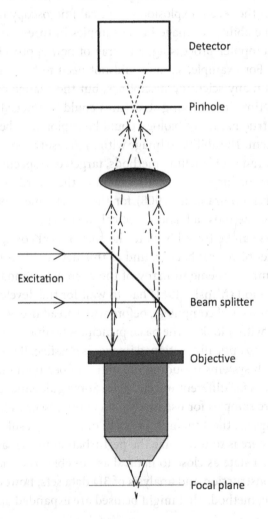

FIGURE 5.1   Basic principles of a confocal microscope. Light from the focal plane passes through a pinhole placed in front of the point detector, while light from outside the focal plane is rejected. The configuration shown here is for an infinity corrected optical system.

the objective lens. Fluorescent light (or backscattered light in some configurations) from the focal plane (shown as a dotted line) then passes back through the objective, and is transmitted by the beam splitter before being focused down through a pinhole onto the detector. Light from outside the objective's focal plane (shown as a dashed line) passes back through the objective and is also transmitted by the beam splitter. However, as this light is not from the objective's focus, it is not collimated on emerging from the back of the objective and on passing through the lens it is rejected by the pinhole. The majority of this out of focus light therefore fails to reach the detector. The system thus only permits light from the focal plane to reach the detector. By scanning the beam, or the sample, an optical slice can then be built up point by point (pixel by pixel). By moving the focal spot through the sample different depths can be imaged.

The original system, which was aimed at imaging neurons, recorded the image on a long persistence cathode ray tube that was then photographed. The first "user friendly" system was built in 1967 (Egger and Petran 1967) who used the technique, in reflection mode, to image ganglion cells and brain tissue. Building on this concept the theoretical aspects of confocal imaging were explained (Sheppard and Wilson 1981). However, all the systems that had been built had the significant limitation that the imaging optics were fixed and the sample had to be mechanically scanned underneath the objective. Although this produced high-resolution images, clearly this was not a practical solution for any sample with dynamic features. The major breakthrough then came with the design of the beam scanned confocal system and the development of low cost personal computers (White and Amos 1987; White, Amos and Fordham 1987). This method, described below, with various optical variations, has now become the technique used in all conventional beam scanned optical microscopes.

5.2 BEAM SCANNED CONFOCAL SYSTEM

The core optics of a beam scanned microscope are shown in Figure 5.2. Light from the excitation source (almost universally a laser, as a high brightness, good quality light beam is required) is directed onto the sample using a dichromatic beam splitter. This can be replaced with a partially reflecting optic for reflection imaging. The light then passes onto a pair of scanning galvanometer mirrors before being directed onto the back aperture of the microscope objective. Optically the important feature here is that the beam coming off the first mirror must not move significantly on the surface of the second scanning mirror. This can be achieved in two standard configurations now used in all beam scanned confocal microscopes. The first and most commonly used configuration is that of *close-coupled* scanners. The alternative configuration, and that illustrated in Figure 5.2a, is known as a *re-imaged* or *4-F* system.

In the close coupled system, the orthogonal mirrors are mounted as close as possible to each other so that when the beam is reflected off the first mirror even at its maximum scanning angle it does not move far on the second mirror. This configuration has the advantage of being compact and requires the minimum number of optical components. However, the

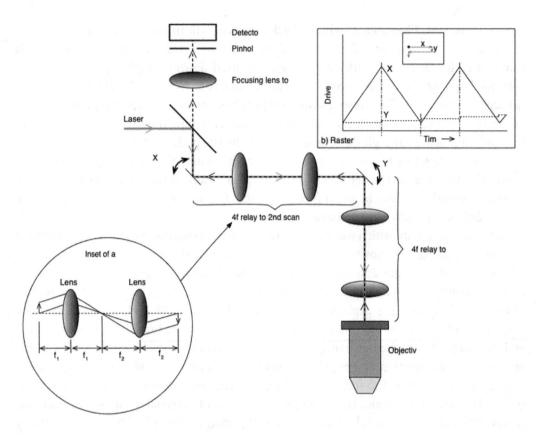

FIGURE 5.2   Basic beam scanned confocal microscope system, a) shows the optical configuration with the inset shows the position for lenses in a 4-F system, b) illustrates raster scanning.

beam does move slightly on the second mirror and when the scanning system is eventually relayed onto the back aperture of the imaging objective there is a small level of lateral beam movement, which can cause beam clipping if the system is not perfectly aligned. In all practical systems this small beam movement does not noticeably affect the imaging quality.

In the alternative configuration, *re-imaged* or *4-F*, the scanners are re-imaged onto each other using lenses, as illustrated in Figure 5.2a. This 4-F methodology is commonly used in optics to ensure that as a beam is moved in one plane it is the angle, not its position, that moves in the other. Optically, for precise beam scanning this is the preferred solution, but the lenses, or mirrors, need to be placed at their focal distances apart as shown in the inset in Figure 5.2. Even with simple, fairly short focal length lenses, of say 100 mm, this means that the two planes being relayed will be 400 mm apart. In the original beam scanned confocal design concave mirrors were used, which meant that the optical path can be folded to reduce the space taken in the scan head (Amos 1994).

In most beam scanned systems, the scanning mirrors are conventional galvanometer operated elements in which the mirror is attached to a rod that is rotated by an applied electric voltage: in essence a miniature electric motor that only rotates over a small angle. Such scanners can be operated dynamically (i.e. moving all the time) or "parked" in a specific position by fixing the voltage, rather than applying a "saw-tooth" pattern to move the mirror. The sample is then scanned in a raster manner as shown in the top right of Figure 5.2. The system can either record data in both X directions of travel (bi-directional scanning) as illustrated, or only in one direction (unidirectional scanning). The former is faster as time is not wasted on the "fly back" to start the next horizontal scan. Initially this time was used to move the data around in the computer but with modern electronics this is no longer required and thus most systems now operate in the bi-directional mode.

Basic X-Y raster scanning systems are now available with scan speeds across the full field of view at up to 30 frames per second for a 256×256 pixel image (for a 2 kHz scanner). Operating for extended periods of time at this speed can impose a strain on the scanner electromechanics leading to heating in the system. Thus higher speed scanning is sometimes undertaken with a more restricted field of view if images are going to be recorded for extended periods of time. It is also possible to only scan certain regions of the image in so-called *region of interest scanning* which means that only these regions are scanned, increasing the potential imaging speed.

In systems requiring high speed imaging one of the scanners (X in Figure 5.2) can be replaced with a resonant scanner. This is a galvanometric device in which the rotating component has a resonant frequency (typically between 4 and 16 kHz) when driven with a sine wave. Here the scanner will move over a large angle very quickly but the movement of the beam coming off the mirror is not perfectly linear in time. However, over a limited region the scan can be considered linear (in effect the beam dwell time at each pixel is uniform). If data is only recorded in this region of the scan, uniform imaging is possible. The other disadvantage of resonant scanning is that the exact position of the mirror is not known from the drive voltage. Thus, either an extra feedback system has to be added to give the mirror position, perhaps using a rotary encoder, or the position of the mirror relative to the drive voltage is calibrated and a look-up table used to convert the drive voltage to the actual beam position in the sample. In addition, the scanner cannot be parked at a particular position, as clearly a constantly applied voltage will not drive the system at resonance. Therefore, although such systems have been built, and sold commercially, they are not common and only used for applications requiring high-speed confocal imaging.

System variations are also possible in which the beam scanning is undertaken using acousto-optic modulators. These are optical devices in which a high frequency sound wave (typically around 40 MHz) is sent across a crystal to produce a Bragg diffraction grating. As the sound wave travels across the crystal the output beam is scanned as the diffraction angle changes due to the sound wave in the crystal at very high frequency permitting high speed scanning. The major complication, ignoring expense and drive challenges, is that the device is then spectrally sensitive as different wavelengths are diffracted at different angles. Such devices are also not suitable for scanning the ultra-short pulses used in non-linear microscopy, as explained in Chapter 8, unless special precautions are taken as they cause the laser pulses to lengthen in time.

With all confocal scanning systems, the returned light is re-imaged by the objective onto the scanning mirrors so that the beam is descanned to subsequently pass through the dichromatic mirror and the pinhole, to reach the detector. Without the descanning on the return path the beam would only pass through the pinhole in one position of the scan. The descanning means that only a single pinhole is required rather than one for each position of the scan. The multiple pinhole method is used in Nipkow disk scanning systems described in Section 5.5. The actual size selected for the pinhole in a confocal system is an important parameter to consider and this is discussed in Section 5.6.

In many systems a series of dichromatic mirrors is used with multiple detectors enabling the fluorescence from a number of fluorophores to be collected simultaneously. The considerations on filter choice, or even the use of a spectrometer before the detector, are provided in Section 5.3. In Figure 5.1 the optics will work both for fluorescence and reflection imaging; the latter is an underused method in complete biological samples as the reflection image can help to place the fluorescence image in context. This is even more valid when an infrared light source is used with its greater penetration and reduced scattering.

The detected signal for each point in the scan is digitized and stored in a computer and the image thus built up pixel by pixel. After a scan has been obtained for one plane the objective can be moved towards or away from the sample to build a full 3D image of the specimen. Typically, a single image plane can be recorded in around 0.5 seconds for a 512×512 pixel image, Higher speeds are clearly possible but at a lower number of pixels per scan and a reduced field of view. Details on the considerations on the scan settings such as the speed and pixel dwell time are described in Section 5.6.

A standard confocal commercial imaging system is shown in Figure 5.3. Due to the requirement for the system to be stable during a scan the best practice is to mount confocal imaging systems, and indeed all high performance optical microscopes, on a vibration isolation system. This prevents vibrations from the building being transmitted into the sample. Such vibrations can cause movement of the sample and optics relative to each other leading to blurring in the image. As many systems are not placed on the ground floor or basements of buildings, where vibration is generally minimized, investment in a vibration isolation system is highly recommended. It may add about 3% to the cost of

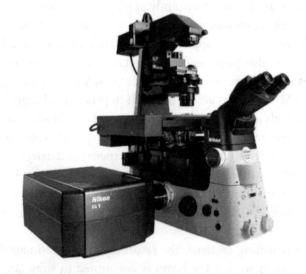

FIGURE 5.3   A commercial scanning confocal microscope (courtesy of Nikon UK Ltd.).

a confocal system but will rapidly pay back this investment in reduced frustration and improved images.

5.3 FILTER SELECTION FOR BEAM SCANNED CONFOCAL SYSTEMS

A crucial component that is perhaps not considered closely enough in a beam scanned system is the choice of the dichromatic filters used in the signal path. These are highly complex structures composed of a glass substrate coated with multiple layers of dielectric materials all a specific fraction of a wavelength thick. The main beam splitter, in a fluorescent confocal system, is designed to reflect the excitation wavelength (or multiple lengths in most systems) with a series of very narrow wavelength band reflectors. The reflectance and transmittance of a typical visible element is illustrated in Figure 5.4. Between three reflective wavelength bands the element acts as a highly transmissive window. This means that the fluorescent photons pass towards the detector with minimal loss. However, any reflected, or backscattered, excitation light is rejected as it is reflected back towards the source.

In general the beam splitter should be optimized for light transmission as one normally has "spare" laser power in the system. It is the fluorescent photons that are precious and need to be preserved. If some laser light is still detected in the signal channels this can be

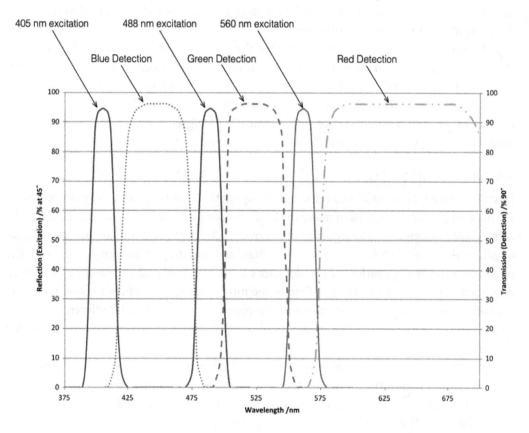

FIGURE 5.4 Typical multi-chromatic optical element in a visible, three excitation line confocal microscope.

removed by incorporating a narrow wavelength rejection filter (for the laser wavelength) directly before the detector.

More recently several commercial systems have been developed that incorporate a spectrometer (or complex system of multiple wavelength dichromatic beam splitters) to increase the spectral sensitivity. These do provide extra information on the emission spectrum detected, but at a loss of optical efficiency. In certain circumstances, where multiple fluorophores are being used, or where a slight change in the spectral profile of a fluorophore is of interest, the loss of detection efficiency can be justified. However, these are exceptional rather than routine requirements for most applications.

In commercial systems the filters are typically matched to specific classes of fluorophores. However, full transmission curves for the filters should be examined and cross-referenced to the fluorophore's published emission. If one is imaging a particularly light sensitive sample (either because the fluorophore bleaches or the sample can be damaged by high excitation light) then it may be sensible to contact the microscope supplier, or a filter manufacturer, to request a coating that is specifically optimized for the sample one wishes to image. The extra cost may well make the difference between successful, minimal perturbation imaging and a bleached and unhappy sample. These filters should also, for obvious reasons, be kept clean and scratch free. Grease on the surface, or scratches, will not only increase the scattering of the precious fluorescent photons, but also can lead to transmission of unwanted wavelengths of light.

After the main dichromatic beam splitter the different wavelengths are directed into the detectors. Due to the physical design constraints on such optical elements the shorter wavelength light is reflected and the longer wavelengths allowed to pass through the dichromatic mirror. Thus in a typical fluorescence confocal the first detector is for the ultraviolet, or blue, excited fluorescence (typically blue/green wavelength emission), followed by the yellow detection and finally the longer wavelengths.

5.4 DETECTOR SELECTION FOR BEAM SCANNED CONFOCAL SYSTEMS

Having scanned the sample, and returned the signal through the filters and pinhole, one of the most important decisions to make in the design, or purchase, of a confocal microscope is the choice of detector. The signal photons need to be converted into an electrical signal as efficiently as possible so that they can be digitized and used to build up the image. The physics behind the operation of the detectors used in a confocal microscope have been described in Chapter 2 Section 2.6. One of the most important features to consider in the case of confocal imaging is the choice of the photocathode material. Different materials have a range of sensitivity to the spectrum of wavelengths of incoming light as illustrated in Figure 5.5.

Generally in a confocal microscope the emission is in the visible and near infrared, typically in the region from around 450 to 750 nm. For the shorter wavelength regions, the choice is typically between the *Multialkali* or *Bialkali*. The Bialkali (materials such as SbKCs) are well suited for use in the blue/green portion of the spectrum, up to around 500/520 nm, though they are still reasonably sensitive to around 600 nm. Multialkali (materials such as $SbNa_2KCs$) have similar sensitivity in the blue but significantly higher

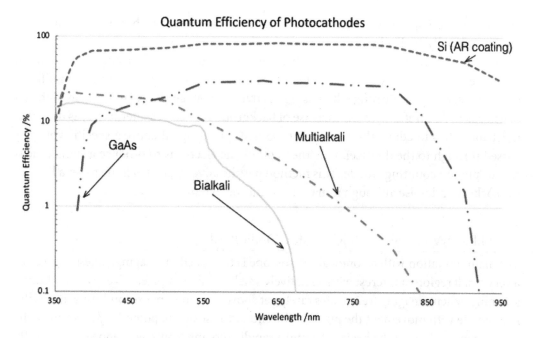

FIGURE 5.5   Photocathode sensitivities for visible and near infrared confocal imaging.

sensitivity towards the red end of the spectrum. However, this greater sensitivity comes at a higher background noise level due to there being more thermally excited electrons to produce a signal (see Chapter 2 Section 2.6.2). Therefore, ideally a Bialkali would be the choice for the short wavelength detection and the Multialkali for the middle of the optical spectrum. For detection of fluorophores above 650 nm the Multialkali sensitivity decreases and the preference moves towards GaAsCs detectors; again care needs to be taken to compare the signal level to the expected background noise level.

Hybrid photomultipliers (Chapter 2 Section 2.6.3) have similar detection wavelength characteristics, as would be expected as they have the same photocathode materials, but generally they have a lower dynamic range than conventional photomultipliers. The early hybrid detectors also suffered from premature failure if exposed to high light levels when active, though faster detection electronics have reduced the risk of damage by switching the detector off quickly in the case of excessive light. These hybrid detectors are significantly smaller than conventional photomultipliers and can be found in an array format. They are now available in some systems to improve the resolution as they can act as their own pinhole either using the light just from the central "pixel" or using a larger area.

There is an increasing use of silicon-based single pixel detectors running in the direct avalanche mode. These have a slightly higher sensitivity than photomultipliers but have a significantly smaller active area (a few microns compared to several millimeters) but are generally more noisy. However, they are more robust, lower in cost and only need a single 5 V power supply with no high voltage present anywhere in the system. Single photon avalanche photodiodes (Chapter 2 Section 2.6.4) do offer the potential for miniature arrays of detectors, perhaps directly after a spectrometer, but at present such systems are not readily commercially available.

The output from the detector can subsequently be amplified electronically, if required, before being digitized. Most confocal systems now use a separate electronic amplifier to provide greater dynamic range than using the dynode voltage alone. However, at high gain one needs to be careful with the increase of background noise. An alternative, as described in Chapter 2, is to adjust threshold limits so that only a signal above a specific voltage is passed through to the digitizer. One then counts the electrons to produce a signal, the so called "photon counting" mode. This method has the advantage of helping to set a limit on the background noise although it may also limit the dynamic range.

5.5 NIPKOW OR SPINNING DISK CONFOCAL SYSTEMS

The main limitation with a confocal microscope is the speed of imaging, unless one selects a very small region of interest with a relatively small number of pixels. Even with a resonant scanning system imaging frame rates rarely get above 40 frames per second. This means that in a sample with movement the pixels at the end of the scan will potentially be seeing a different scene to those at the beginning, and a rapidly moving feature will appear blurred. The pinhole concept, however, does provide good optical sectioning, and indeed even improves the lateral resolution by a factor of around $\sqrt{2}$. A variation on the beam scanned confocal microscope is through the use of a Nipkow spinning disk. This was in fact the scanning technique used in the original John Logie Baird television system! The configuration for use as a confocal microscope is illustrated in Figure 5.6. An expanded laser beam is initially incident on a rapidly rotating array of micro-lenses. These focused "beamlets" are directed at a matching array of holes onto the objective lens and subsequently onto the sample. The returned light then passes back up the optical system passing through the same hole and is reflected back onto a camera (or in fact the eye can be used). By careful arrangement of the lenses in a spiral pattern it is possible to scan the entire sample at high speed. Typically, over 7000 lenses and holes will be used rotating at speeds of 1000 rpm.

These systems provide some optical sectioning capability but are not as efficient at rejecting the out of focus light as the more conventional point scanned system (Egner, Andresen and Hell 2002). In addition, they are wasteful of the laser light as only a small portion of the beam is used at any one time. In more scattering samples the emission light can be scattered to adjacent pixels in the camera leading to a significant loss of contrast which will further decrease rapidly with depth. The main advantage of such a system is that very rapid imaging is possible with frame rates well in excess of 100 fps. The limitation is mainly due to the sensitivity of the camera and fluorescence efficiency of the sample. Spinning disk confocal microscopes are well suited to imaging calcium transients, microtubule dynamics and fluorescence from highly punctate samples at high speed and are commercially available from several sources. With the advances in light sheet, or single plane microscopy (Chapter 7), the use of the spinning disk is decreasing but due to its conventional epi-configuration it is still an optical sectioning imaging method to be considered for many samples.

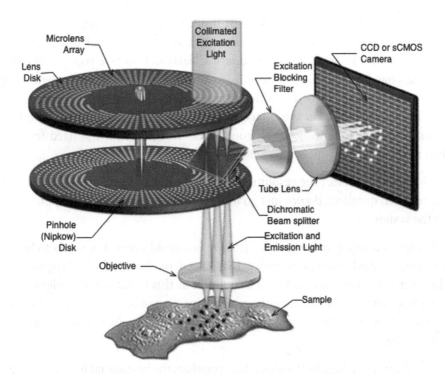

FIGURE 5.6 A Nipkow or spinning disk confocal microscope (Based upon image credited to Molecular Expressions.com at Florida State University).

5.6 PRACTICAL GUIDELINES TO MAXIMIZE THE PERFORMANCE OF A CONFOCAL MICROSCOPE

Having outlined the basic physical principles behind a beam scanned confocal microscope, the challenge is to obtain the images, or data sets, required to understand the processes and structure of the sample under investigation. Before considering how the various parameters of a confocal microscope may be best adjusted for maximum performance one should consider carefully what information is required from the sample. Clearly one would like the best possible image but this should not be the overriding principle in recording the data. The main reason for using a confocal microscope is to obtain 3D information without physically sectioning the sample. One may also want to observe dynamic changes taking place. Thus before starting any imaging session one should consider how the resulting data is going to be used, and crucially how you can minimize potential damage to tissue, particularly when imaging live samples. The questions below provide a guideline to the items that should be considered before starting an imaging session. They assume that prior to approaching the microscope one already knows what fluorophores are present within the sample and thus which are the best wavelengths to use both for excitation and detection.

1. *What spatial resolution is required?* Although everyone would like as high a spatial resolution as possible this can lead to slower imaging, more light on the sample, very large data sets and one might have to compromise the field of view. The important

factor to remember here is the *Nyquist* limit. Simply stated: you should image with a spacing at twice the resolution you require. For example, a 1 μm axial resolution requires optical slices to be 0.5 μm apart. Clearly in this case you would also need a lens that has 0.5 μm axial resolution (as described below). Imaging with a spacing much smaller than this will not improve the actual resolution of the images though may reduce the noise on the images. One may therefore decide to undertake such "over sampling" for the "journal front cover" image even if not used for the actual data analysis.

Guideline 1   Select the resolution required to provide the data needed to answer the research question, do not just go for the highest resolution that might be possible on the system.

2. *What speed of imaging is required?* If there is a rapid event that needs to be recorded one would ideally image at twice that speed based upon the Nyquist criterion. However, do not just image as fast as possible as this tends to mean higher excitation powers to obtain sufficient signal per pixel, increasing the risk of phototoxic damage or photobleaching. More rapid imaging also tends to produce images with lower signal to noise.

Guideline 2   Image at the speed that provides the best signal to noise and captures any transient events.

3. *What field of view is required?* This is clearly linked with questions 1 and 2 as increasing the field will increase the imaging time, and may affect the possible resolution. This also relates to the depth of imaging required. In general signal is lost, along with contrast, as one images more deeply. The depth may well help determine many of the settings in terms of laser power and detector sensitivity, and one can start to have saturation challenges nearer the surface in a very deep scan.

Guideline 3   Only image the volume that is required to provide the data needed to answer the question. Do not image large depths just because you can!

4. *What period of imaging is required?* If one is trying to follow a slow process (for example cell movement during wound healing) know how long it will take for the cells to close the wound. Image for sufficient time to observe the effects one is interested in.

Guideline 4   Only image for the minimum time needed to obtain the data set required.

5. *What imaging interval is required?* For example, if one is trying to track cell movement, how close together in time do the imaging datasets need to be? One clearly needs to know how each feature has moved between imaging sets, but again, imaging more frequently increases the light dose to the sample.

Guideline 5   Image as little as possible, but with the time interval between images determined to ensure you know what movement has taken place.

6. *Is the sample particularly sensitive?* This might mean the fluorophore photobleaches easily, the sample is easily damaged by light, or that local changes in temperature, pH or gas concentration affect viability. Is time critical? There may only be a short time to determine the operating parameters and therefore it might be advisable to use a surrogate model to establish the imaging parameters.

   Guideline 6 Image with as low a power as possible, as infrequently as possible. Minimize the perturbation to the sample.

7. *Are further tests going to be undertaken on the sample after imaging?* One might need to know exactly where one has imaged within the sample for a subsequent procedure. This could be as simple as not using an oil objective as the oil may cause a complication for a subsequent process.

   Guideline 7 Plan ahead!

8. *How is the data going to be processed after imaging?* Although this may not be known in detail at the start of a new project, some consideration should be given to this part of the imaging processes. If 3D data stacks are to be recorded for an extended period of time, the data files will be very large and moving them to the processing computer may be an issue. This could determine the imaging protocol, perhaps taking a series of single data stacks rather than a complete time series. The post-processing may also influence some of the settings used to control the laser power, real-time averaging, black level and detector gain.

9. *Is there a feature in the image that can be examined quickly (even just by eye) to confirm the data set is good?* This may reduce the chance of recording data which is not fit for purpose.

   Guideline 9 Do have a quick look at the data before leaving the microscope.

10. *Is there a previous protocol that should be followed?* This may be from a similar experiment, or just based upon experience of using either that sample or fluorophore on a previous occasion.

In summary; plan ahead, minimizing the light exposure to the sample.

5.6.1 Microscope Choice, Sample Mounting and Preparation

The aim of this section is to provide the user with some areas that should be considered in preparing, mounting and imaging a sample in a confocal microscope. It is not intended to replace protocols that have already been developed, but should provide some indication of ideas and directions to explore before imaging a new sample.

Probably the most important consideration here is the type of microscope to be used: inverted or upright, and this may well be determined by the instruments available. Inverted systems are generally more common. This position is changing slightly with the increasing use of non-linear microscopy (which is frequently undertaken on a core confocal instrument) and with the sample frequently alive, or at least more intact. If there is a choice, inverted microscopes are probably the preferred option for imaging cellular layers and thin samples. Usually the sample (unless solid, such as rock) will lie flat on a coverslip and through cellular adhesion and gravity will be stable during imaging. Inverted microscopes also have more imaging "ports" available and this then provides the opportunity to record conventional widefield images as well as the confocal data sets. This can sometimes help with image recognition when only a small area is recorded with the confocal system.

Most objectives that are designed to image through a coverslip (which is the vast majority of lenses) are optimized to work with a 1½-thickness coverslip (around 175 μm thick). Using glass of different thicknesses will affect the ultimate optical resolution but the effect is small, and generally lost in other aberrations as soon as one images a short way into the sample. Due to the weight of the sample, or perhaps because one is using an oil lens and rapid movement causes the glass to bend, a thicker coverslip may be advisable. If the sample is in liquid and additional fluid is required during the imaging processes (such as an agonist), a coverslip bottomed dish can be used. These can easily be made by cutting a hole in the bottom of a plastic dish and attaching a coverslip using a waterproof and non-toxic glue. It should be remembered that a sample could move as a result of the addition of fluids and should therefore be securely fixed. If the sample is mounted in fluid, and going to be imaged for extended periods of time, it is important to cover the sample to prevent evaporation or ingression of dust. It is also important to remember that in taking a 3D data set the objective will advance towards the sample to take the deeper images. Therefore, the mounting of the sample, including the thickness of the coverslip, should always enable the lens to travel towards the sample without hitting anything!

If an upright system is used the great temptation is just to place the sample under the microscope and image away. As mentioned above, most lenses are designed to have a coverslip in the optical path and thus if one does not put one on then spatial resolution will be lost, but the effect is not large. Normally access to the sample is easier with an upright microscope and there is no worry about the lens hitting the area around the sample, just the sample itself! Again some thought should be given to keeping the sample secure as any addition to the sample can cause movement. An advantage of an upright system is the use of water dipping lenses. These generally have a long working distance and a high NA, although not as great as oil immersion lenses. As many biological samples are mounted in water this provides a route to imaging the sample with minimal perturbation. One disadvantage of the upright system is that the top surface of the sample, through which you image, may not be flat and therefore may not be a good reference surface.

In all cases it is essential to ensure that the sample is stable and will not move during extended imaging periods. In prepared and fixed samples this is clearly not a problem but as the confocal is frequently used for more intact samples with vertical, as well as horizontal, extent, sample movement is more of an issue. Even if the confocal is mounted on an isolation platform, vibrations from pumps and equipment fans on the same table as the microscope can cause imaging artefacts and blurring.

Due to the confocal microscope's inherent optical sectioning the sample may not need to be prepared in exactly the same manner as that used for widefield imaging. Prepared slices may not need to be as thinly cut, though it should always be remembered that as one images more deeply the signal and resolution will fall. Increasing the thickness of a physical section can help to ensure that one is looking at the true 3D structure, rather than a distorted version caused by the sample preparation. In certain circumstances this might lead to a change in the way that a sample is dissected to ensure that sound structure is being viewed.

5.6.2 Lens Selection

The lens is probably the single most important component in a confocal microscope. It is also the most expensive, and therefore selecting the most suitable lens for a specific imaging task is vitally important. Everyone has their own favourite lens but when moving to the confocal this may no longer be the best choice. In a confocal system the resolution is set by the numerical aperture of the lens, while the main role of the lens magnification is determining the field of view. The combination of the magnification and NA affects the working distance of the objective lens and when one is looking at thicker samples this clearly has to be considered more carefully than in a sample consisting of a single layer of cells.

The first consideration is the type of lens to use. As most confocal microscopy is undertaken in the fluorescence mode the chromatic properties of the lens are important. This can be appreciated if one considers a sample that is excited at 488 nm and then emits at 550 nm. If using a lens with a numerical aperture of 0.75 there is an axial resolution of around 1 μm. For an achromatic lens, colour corrected for blue and red light only and with spherical correction only in the green (Chapter 3 Section 3.4), the separation between the blue and green planes may be up to 3 μm. This means that the peak of the excitation could be up to 3 μm away from the maximum light collection point and thus much of the fluorescence may be rejected by the confocal system. Therefore the ideal choice is an apochromatic lens (corrected deep blue through to red). This is the type of lens normally used for fluorescence microscopy. In the case of confocal imaging though it is even more important to have the full colour correction.

The second consideration is to use a plan-apochromat. This will ensure a flatter field of view as the beam is scanned over the back aperture of the objective. Although not as critical as selecting the apochromat option, plan lenses do offer significantly better confocal imaging. As all confocal systems operate in the epi-configuration, the higher the numerical aperture of the lens the better the resolution, but it also enables one to collect as many of the fluorescent photons as possible. As fluorescence is emitted in all directions the larger the collection angle of the lens the greater the quantity of light collected. As a rule of thumb there is little point in using a confocal microscope with a lens with an NA of less than 0.75.

The use of a water-dipping lens is of course only possible with an upright microscope and will depend on the sample mounting method that has been adopted. In practice it is often worth using an air lens of around 0.75 NA to have a look through the sample, particularly if it is a preparation that one has not used before. One can then switch to high NA lenses as needed for the resolution required.

As mentioned above, the magnification mainly sets the field of view and one can use the scan zoom on the scan head to give a high magnification of an area. The crucial factor to remember is that to fully utilize the resolving power of the lens the pixels need to be half the size of the resolution of the lens, set by the NA. For example, for a system operating at 500 nm with a lens with an NA of 1.35 the diffraction limit is 0.22 μm. This means that in zooming ideally each pixel would represent 0.11 μm on the sample. For a 512×512 pixel image this would give a field of view of around 57 μm. Zooming in further than this will not increase the resolution, despite the fact that the confocal software might say each pixel is as small as 20 nm! They are not, they are only the diffraction limited size.

In summary, where possible select an apochromatic lens, with a minimum NA of 0.75 and a magnification above ×20. The preference is to use a plan-apochromat lens to help produce more even illumination and collection across the entire field of view.

5.6.3 Initial Image Capture

As has been mentioned frequently, when imaging, one aims to minimize the perturbation to the sample and therefore the target is to use the lowest light possible to produce images with good enough signal to noise for the subsequent image processing. The guidelines below are not intended for any specific commercial, or home built, confocal system. They are aimed at providing a less experienced user with a route to obtaining good quality images. The reason for each step is explained and some users may alter the order of some of the steps based upon personal experience and preference. The guidelines are written for fluorescent confocal imaging but may be modified as required to work for reflection imaging. They also assume that the confocal system has been switched on and all local safety rules have been followed. The preference is to work in a laboratory with subdued lighting in order to minimize the stray light that might reach the detectors.

1. If a conventional fluorescence microscope is available, it is useful to check that the sample is fluorescent before attempting to image using the confocal system. Most commercial instruments include a conventional epi-fluorescence excitation and detection system and this is a worthwhile investment in any confocal microscope purchase.

2. Switch off, or block, the epi-illumination and adjust any filters and shutters required to go into the confocal mode. All experienced users will at one time or another have forgotten to adjust the filter positions and spent time trying to work out why the confocal image is blank!

3. Select the laser wavelengths for the fluorophore of interest (in multiple labelled samples it is worth starting with just one wavelength) and adjust to a low power. The actual setting will depend on experience with a specific system but 10% is a good initial guideline figure.

4. Open the pinhole. This means that one will not have good optical sectioning but the main target is to obtain an image, which can subsequently be adjusted to provide a high quality dataset. Opening the pinhole maximizes the probability that at least some part of the sample will be within the focal range of the microscope.

5. Select the correct detector for the fluorophore in use and adjust the gain to a reasonably low level. Each system will have a different way of setting this but around halfway to the maximum gain is a good compromise.

6. Check that the lens in use is shown in the software. This is important as this parameter will be saved in the image meta-data and also will display the size of the field of view and effective pixel size in the image. Many systems also automatically adjust the pinhole to match the lens for optimal sectioning.

7. Select a scan of around 512×512 pixels and remove any scan zoom (to obtain the entire field of view). Initially imaging at around 1–2 frames per second is a sensible starting speed.

8. Activate the scan.

9. If nothing is seen increase the gain on the detectors to around 80% if required before slowly increasing the excitation power.

10. If nothing is seen at up to around 50% laser power then adjust the focus and check to make sure that all the filters and shutters are in the correct position before considering increasing the power or detector gain.

11. Once an image has been obtained the focus can be adjusted so that the area of main interest is found. One is now in a position to adjust the pinhole size, detector gain and excitation. This can be undertaken using any type of false colour "look-up table" on the image, but one which makes saturation and no signal very obvious is preferred. It is suggested that the pinhole is closed to about halfway and then the gain and laser power adjusted to give a suitable image before shutting the pinhole to the optimal setting for the lens selected.

12. Adjust the detector and laser power for an optimal image (see below for some guidelines) again using a look-up table, which maximizes the extremes of the dynamic range.

13. If multiple fluorophores are being used change the excitation source and repeat the process as required.

14. If a 3D scan is required then select the required options and manually, using one of the wavelengths, observe the top and bottom of the image stack to ensure it is covering the correct volume. At this stage it is also worth checking that the image is not saturated on a specific image plane, or group of planes.

15. Record the image stack, and make a note in a lab book, or preferably a computer text file or spreadsheet, of the file name and other conditions that may not be captured in the imaging meta-data. The routine use of a computer-linked file for every imaging session is suggested as this can make it easier to search for a particular image set later.

16. Before moving onto the next sample it is worth having a quick look at the recorded data set to ensure it is fit for purpose.

17. At the end of the imaging session copy the files recorded and ensure that they are backed up! The local protocols may also suggest that you then remove the data and temporary files from the instrumentation computer so there is plenty of storage space for the next user.

5.6.4 Optimization

As has been mentioned several times in this chapter, and is a recurring theme throughout the book, the ethos of all imaging is to obtain the best images that are suitable for the task in hand with the minimum perturbation to the sample. In the case of confocal microscopy this could be summarized as "keep the laser power low". If one is only using 0.1 mW of laser power with a 1.35 NA objective lens (a typical oil objective) at 500 nm then the focused spot intensity is around 250 kWcm^2! Even though a single point may only be scanned for 10 µS this can clearly be enough to cause damage to the sample. This is even before consideration is given to photobleaching of the fluorophore.

The following notes provide a few guidelines on optimizing and imaging once something is visible on the computer screen. It is also suggested that a look-up table, in which low values and high values appear as colours on an otherwise black and white image, is a good way of testing the system. Figure 5.7 demonstrates a few of the imaging errors discussed below.

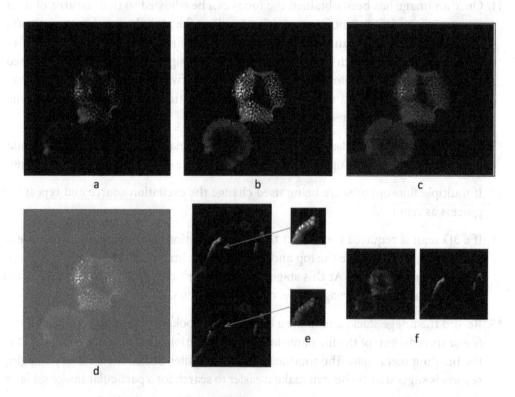

FIGURE 5.7 Confocal images illustrating different settings and faults: a) optimal settings, b) gain too high with saturated pixels, c) black level set too high with missing features, d) insufficient gain or signal, e) series of images showing photobleaching, f) single image and Kalman averaged image.

1. *Gain and Black Level Setting* With most modern detectors it is possible to adjust the gain and the black level of the detector separately. It is worthwhile experimenting with these features on a non-important sample at very low laser powers. Even something as simple as a dilute solution of fluorescein in water is suitable (and is useful for alignment checks discussed later in this section). If some non-fluorescent features can be included as well (non-fluorescent beads for example which will fall to the bottom of the sample under gravity) then one also has some areas of zero intensity. Alternative simple test samples do include fluorescent beads as a weak water suspension or agar gel, or fluorescent plastic slides (though these can photobleach with repeated use in the same area).

   With such a sample, at very low excitation intensities, it is then possible to observe the effect of increasing the detector gain (while keeping the black level constant). As the gain is increased the signal will increase, but at a certain value (if using a conventional black and white image) white spots will start to appear over the image. This is due to noise generated in the system being amplified and starting to affect the image. The rule is to reduce the detector gain from this point so that high contrast images can be obtained without this noise.

   As one adjusts the black level one is effectively cutting off the lower values in the image. If one sets the black level too high then one will be removing real, but weakly fluorescing features. The aim, once the gain has been set, is to adjust the black level so that the black areas of the image, where there is known to be no fluorescence, do become black.

2. *Image Averaging* Providing the sample is not photobleaching rapidly, and is temporally stable, image averaging provides an excellent way to increase the image quality. The typical filter used is a Kalman filter, which works in real time. The mathematical background is beyond this book but can be found in books and web pages. The important point for our consideration is how many images need to be averaged to give a reliable result? Experience says between three and four images provides a good balance between time (and hence potential sample damage) and good signal to noise.

3. *Assessing Image Quality* There are multiple metrics that can be used for assessing the quality of an image and this is a complex field in its own right, but probably the simplest is to use a histogram. This simply plots the number of pixels with a given intensity against the pixel intensity value. Examples for a "good" and "bad" image are given in Figure 5.8a, b. In the good image it can be seen that there is a good spread of pixel values illustrating that the image has high contrast. Whereas, for the bad image the pixels are all centred around one value. This indicates an image with little contrast, making features hard to see as can be observed in the accompanying image. Many systems now provide a real time histogram and the various settings on the instrument can then be adjusted in an attempt to produce a well-spaced histogram. Ideally one wishes to use the entire dynamic range of the system but many systems can become non-linear near the high value end and thus it is preferable not to have many saturated pixels.

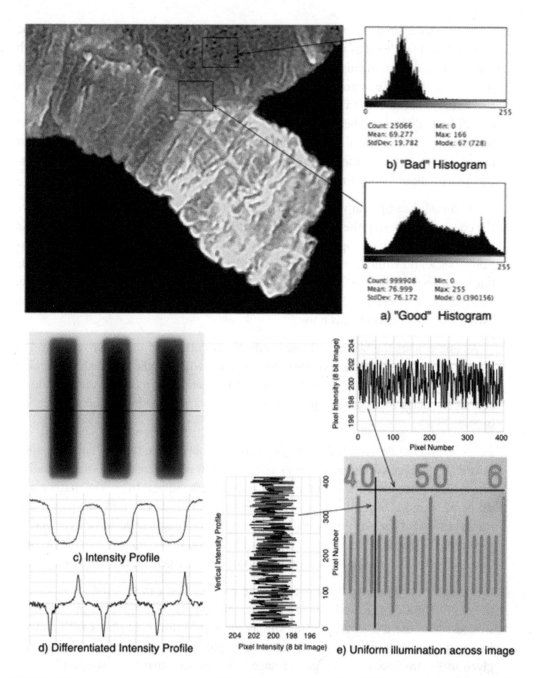

b) "Bad" Histogram

Count: 25066   Min: 0
Mean: 69.277   Max: 166
StdDev: 19.782   Mode: 67 (728)

a) "Good" Histogram

Count: 999908   Min: 0
Mean: 76.999   Max: 255
StdDev: 76.172   Mode: 0 (390156)

c) Intensity Profile

d) Differentiated Intensity Profile

e) Uniform illumination across image

FIGURE 5.8 Assessment of image quality: a) image with a "good" histogram, b) image with a "bad" histogram, c) line profile through a test target using a fluorescent sea and dark bands, d) differentiation of line profile to provide a measure of the image resolution, e) optimal illumination in a confocal system.

4. *Assessing Resolution* There are two methods that can be used to obtain a measurement of the optical resolution of a system. Perhaps the easiest is to use fluorescent beads of a known size, and ideally less than the diffraction limit of the system. For a confocal with a 1.35 NA oil objective these would typically be 100 nm beads. By imaging the bead and then measuring its size one has an idea of the resolution. As the size measured is a combination of the actual bead size (and there will be a variation from bead to bead) and the point spread function of the instrument actually determining the resolution, this way is not trivial and requires deconvolution of the images and knowledge of the exact bead size. However, as a simple method to see if the system is well aligned, or is changing from day to day this provides a simple test. It should also be noted that as one images more deeply into the sample the size of the bead will appear to increase due to sample induced aberrations (discussed in Chapter 8). One should also be aware that in a thin sample of prepared beads it can be hard to find the beads using the confocal configuration and the use of conventional fluorescence imaging to find the beads is highly recommended.

An alternative approach is to use a sample which has a rapid change from no signal to signal. In the vertical direction this may be a fluorescent sea and just a simple solution of a fluorophore is sufficient (Zucker and Price 2001). In the horizontal direction it may be a test target with specifically marked black areas (USAF test target) or a micro-machined fluorescent plastic (Corbett et al. 2014). In both cases one takes a simple optical slice (or vertical slice in the case of axial resolution). With any image processing software (for example ImageJ or Fuji) one then draws a line through the rapid change in contrast. The intensity profile of the line can then be exported into a spreadsheet and simple differentiation (change in X divided by change in Y for any two adjacent pixels) will provide a good measurement of the resolution. The results of such an approach are shown in Figure 5.8.

A differentiation in a spreadsheet can be accomplished by simply taking the change from one line to the next in the columns of interest. For a spatial column (i.e. the pixel position for the line mentioned above) that starts at A0 and moves down, one would calculate A1–A0, then A2–A1 all the way to the bottom of the column and then repeat this for the intensity column. Then plotting the two result columns against each other will provide the resolution of the system if the width of the resulting curve is measured at the half signal point.

5.6.5 Saving Data

A great deal of time and cost will have gone into the preparation of the sample and imaging and thus saving the datasets and backing them up is crucially important. Most imaging centres have a standard protocol for data saving, transferring files and backing them up but generally these will be the responsibility of the user. Either ensure there is a high-speed network to transfer the files, or take along a suitable computer drive to copy the images for each imaging session. 3D data stacks can become large. For a dataset consisting of

100 planes with three colours one can easily have 300 Mb image files, and if a time sequence is undertaken the files can become very large. The raw data files should always be the ones copied along with any meta-data associated with the image set. Journals, and funding bodies, are increasingly asking for all the data to be saved and made available, therefore saving all this data, carefully, is becoming increasingly important. It is best practice not to actually process the master image file, as there is always a risk of accidentally overwriting the original data. Copying the data sets and working with the copy is always safer.

As suggested in the short imaging guidelines above it is good practice to keep a separate file, or lab book, in which details of the imaging session, and most importantly, the file name are recorded. Using a spreadsheet it is possible to enter the image file name, the sample preparation protocol as well as other useful data such as the lens and image dimensions rapidly and in an easily searchable form. In addition, once the correct headings are set up in a spreadsheet, it is easy to ensure all the relevant details are entered each time. Although these parameters are also found in the instrument meta-data, having rapid access to a previous imaging protocol is frequently a useful place to start when one has a new sample to image.

5.6.6 Routine Maintenance

Although most large imaging centres undertake regular maintenance of confocal microscopes there are a few simple tests that can be undertaken, in particular if the system seems to be producing poor quality images. The most basic method is to take a fluorescent sea and image a short distance into the sea. One should have a uniform image with all the pixels close to the same intensity (Figure 5.8e). If the image is not uniform (and this can clearly be seen using a multicolour look-up table) then the alignment of the laser into the scan head may need adjusting. Generally, there will be a small tail-off in intensity towards the edges of such a fluorescent sea image but this should be uniform. Ideally this measurement should be made at known power and detector settings so any slow degradation in performance in the instrument will be spotted.

One can then, using a sample with a clear feature, image and place the feature in the centre of the field of view. As one then uses the zoom facility of the confocal system the feature should stay in the centre of the field of view. If this does not happen then the scanning system requires adjustment. A further simple test with this type of target is to change lenses; if the lens turret is operating correctly the feature in the centre of the field of view should not alter. If there is significant movement, the lens turret will require adjustment; a problem more likely to occur on upright systems.

It is also good practice to have a slide with a graduated ruler etched into the surface. This is likely to have a 1 mm line with 10 or 100 μm markings. This should be imaged and the spatial calibration of the system checked using the measuring facility that is present in most microscopes. If this is incorrect, the most likely error is that the wrong lens has been selected, or the lens look-up table in the system has become corrupted.

The final basic test is to undertake the resolution measurements described above. Again the absolute value is not critical for most applications, but if such measurements are recorded on a daily basis then a drift in the performance of the system can be detected.

It is also good practice with a microscope that is used by multiple people to have a regular policy on the removal of data files, though clearly all users should be aware of this before their precious data sets are destroyed. Preventive maintenance, both hardware and software, is always better than a catastrophic failure when a valuable sample is about to be imaged.

5.7 RECONSTRUCTION

Having recorded a wonderful XYZ data stack one now has to consider how this will best be used to answer the specific research question of choice. There are multiple software packages available to process the data and these vary in cost from free multi-user shareware to highly expensive single licence software suites. Reconstruction software is also available from most microscope suppliers though this is normally basic and best used for checking the images at the microscope before an imaging session is completed.

ImageJ (https://imagej.nih.gov/ij/download.html) is open access code that is freely available and used by most laboratories around the world. It runs on all computer platforms as it is written in Java and it is possible to develop complex image processing pipelines using the macro-software. As there is a worldwide user base there are also widely accessed user communities and forums. This means that for some initial image processing and reconstruction there is probably a suitable package already readily available.

In undertaking the 3D reconstruction, the final use of the images should be considered. It is now possible to upload video files for most publications, and thus 3D videos which rotate to give a representation of the 3D nature are widely used. However, when undertaking such a reconstruction it should be remembered that the axial resolution will be lower than that found laterally, and thus certain features can become distorted. The colour lookup table that is used can also drastically affect the way an image set is perceived, and careful thought should be given to the final presentation to ensure that the features of interest are clear. The more expensive software packages provide improved methods of rendering the final image set, for example to apply a surface finish to a structure. Such images are undoubtedly of a higher quality than the shareware options, but the packages can be complex to operate and considerable time can be wasted just trying to get the perfect video, which may only be played a few times at a presentation. If the data set contains dynamic data then the more expensive software available generally has to be used.

REFERENCES

Amos, W. B. 1994. "Confocal Scanning Optical Microscope Patent 5,304,810".

Corbett, A. D., R. A. Burton, G. Bub, P. S. Salter, S. Tuohy, M.J. Booth and T. Wilson. 2014. "Quantifying Distortions in Two-Photon Remote Focussing Microscope Images Using a Volumetric Calibration Specimen". *Frontiers in Physiology* 5(October): 384.

Egger, M. D. and M. Petran. 1967. "New Reflected-Light Microscope for Viewing Unstained Brain and Ganglion Cells". *Science* 157(4): 305–7.

Egner, A., V. Andresen and S. W. Hell. 2002. "Comparison of the Axial Resolution of Practical Nipkow-Disk Confocal Fluorescence Microscopy with That of Multifocal Multiphoton Microscopy: Theory and Experiment". *Journal of Microscopy* 206(Pt 1): 24–32. www.ncbi.nlm.nih.gov/pubmed/12000560.

Minsky, M. 1961. "Microscopy Apparatus". www.google.com/patents/US3013467.

Sheppard, C. J. and T. Wilson. 1981. "The Theory of the Direct-View Confocal Microscope". *Journal of Microscopy* 124(2): 107–17.

Shimomura, O., M. Chalfie, W. R. Kenan and R. Y. Tsien. 2008. "The Nobel Prize in Chemistry 2008 'for the Discovery and Development of the Green Fluorescent Protein, GFP'". *Nobelprize.org.*

White, J. G. and W. B. Amos. 1987. "Confocal Microscopy Comes of Age". *Nature* 328(6126): 183–4. http://dx.doi.org/10.1038/328183a0.

White, J. G., W. B. Amos and M. Fordham. 1987. "An Evaluation of Confocal versus Conventional Imaging of Biological Structures by Fluorescence Light Microscopy". *Journal of Cell Biology* 105(1): 41–8.

Zucker, R. M. and O. Price. 2001. "Evaluation of Confocal Microscopy System Performance". *Cytometry* 44(4): 273–94. www.ncbi.nlm.nih.gov/pubmed/11500845.

CHAPTER 6

Fluorescence Lifetime Imaging Microscopy (FLIM)

T<small>HE USE OF THE</small> fluorescence lifetime as an imaging contrast mechanism has grown significantly in the last twenty years, mainly as a result of advances in technology. Lower cost ultra-short pulsed light sources, more sensitive and faster detectors, high-speed electronics and more powerful computers have all helped contribute to the greater application of the method. The technical advances have been supported by improvements in chemistry enabling the location of fluorophores to specific sites within cells alongside the ability to design fluorescence systems to act as Förster resonant energy transfer (FRET) pairs. The technical ability to apply short pulses of light, and to then detect the emission very accurately in time, is also the basis of several super-resolution microscopy methods described in Chapter 12.

This chapter starts by introducing the physical processes involved in the generation of fluorescence and how this links to the fluorescence lifetime of a molecule. It then considers how the lifetime is affected by the local environment before describing the methods used to generate fluorescence lifetime images. At the end some basic guidelines to be considered while using lifetime imaging are discussed alongside some examples where the technique has really demonstrated that it is the method of choice for examining specific processes. In several places reference is made to Chapter 5 on confocal imaging as the majority of lifetime imaging is undertaken through a beam scanned imaging system.

6.1 INTRODUCTION TO FLUORESCENCE LIFETIME

6.1.1 Absorption and Emission

The basic concept of fluorescence is straightforward and is illustrated in Figure 6.1. Light at the correct wavelength is absorbed by a molecule, or atom, causing an electron to move to a higher energy level. The electron then loses a small quantity of energy to its surroundings as it moves to a slightly lower energy level from which it then falls back to the initial

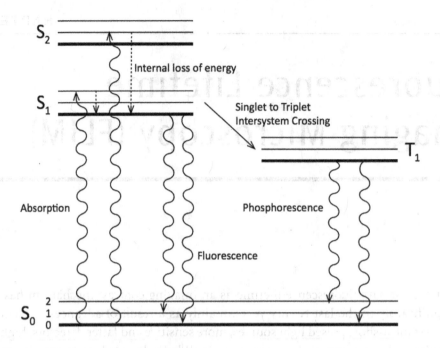

FIGURE 6.1 Jablonski diagram.

starting level with the emission of light. As some of the energy initially absorbed has been transferred to the system the emitted light is at a longer wavelength than that originally incident on the sample. This change in wavelength is known as the "Stokes shift" after the physicist George Stokes. In a typical fluorophore used for imaging, this process takes under 10 nanoseconds and it is the average time taken for a photon to be re-emitted that defines the fluorescence lifetime of a molecule.

One of the first examples reported in academic literature of fluorescence was by Sir William Herschel (1845), normally thought of as an astronomer. He described the blue fluorescence seen in a solution of quinine when it was exposed to the ultraviolet component of sunlight. This is an easy experiment to repeat as quinine is one of the ingredients in tonic water. On the surface of a glass of tonic water exposed to the ultraviolet light of the sun a blue glow can sometimes be seen. The glow is in fact increased by altering the polarity of the solvent, for example with the addition of ethanol (gin working well in this experiment!).

Luminescence is the emission of light from a substance due to changes in an electron's energy. Technically luminescence is divided into two types, *fluorescence* and *phosphorescence* depending on the energy levels involved. With reference to Figure 6.1 the singlet ground and first two excited states are labelled S_0, S_1, S_2. A photon can be absorbed to excite an electron from the singlet ground state to a singlet excited state following the quantum mechanical rules governing allowed transitions (it will be to a state with the opposite electron spin). This electron will rapidly lose some vibrational energy and then decay rapidly back to the ground state energy levels with the emission of the photon. This process is rapid.

However, it is also possible for the excited electron to cross over to an excited triplet state (T_1 in Figure 6.1). According to the quantum mechanical transition rules direct return to the S_0 ground state is forbidden and thus the emission of light via this route is slow with delays of up to several seconds. This is the process known as phosphorescence.

In this process there are two important characteristics for all fluorophores, the fluorescence lifetime and the quantum yield. The quantum yield (Q) is the number of emitted photons relative to the number of absorbed photons. The lifetime (τ) of a fluorescence process is the summation of multiple contributions and beyond the scope of this book but can be found in the standard textbook on fluorescence (Lakowicz 1999). We only need to consider that the natural lifetime (τ_n) is given by $\tau_n = \tau / Q$.

The absorption and emission spectrum of probably the most common fluorophore, fluorescein, is shown in Figure 6.2. As can clearly be seen it has a very broad absorption profile with all the individual electron levels (with the ground state S_0 and first excited state S_1) overlapping due to the thermal energy present in the system. The exact spectrum does change slightly with the environment. Specifically for fluorescein the quantum efficiency is affected by the local pH and in general the spectrum of fluorophores can be affected by the local environment (ion concentration, pH, viscosity, lipid or protein level, etc.). As can be seen in Figure 6.2 for fluorescein there is a high spectral overlap between the absorption (peak ~ 494 nm) and emission (peak ~ 512 nm) spectra, giving a Stokes shift of only around 18 nm. For this reason it is often excited below the peak wavelength at around 460–488 nm

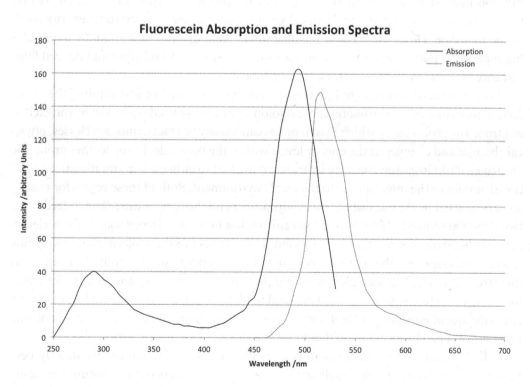

FIGURE 6.2   Absorption and emission spectra of fluorescein around neutral pH.

though this is partly determined by the availability of light sources at these wavelengths. However, it remains a popular fluorescent label as it is highly water-soluble, is non-toxic (it is used in humans for retinal angiography for example) and has an emission wavelength close to the peak of the sensitivity of the human eye.

6.1.2 Fluorescence Lifetime

In terms of using the lifetime of a fluorophore as an enhanced contrast mechanism, and a source of extra information about the environment surrounding the fluorophore, the process after absorption, the loss of energy to the surrounding environment, is crucially important. In one extreme the emission can be removed through processes known as quenching. Quenching can occur through multiple means and the most likely is collision with another molecule that can absorb the energy present in the excited state. The excited molecule thus returns to the ground state without emitting a photon. Molecules that act as quenchers under certain circumstances include oxygen, halogens and amines. We will return to this process in Section 6.4 when discussing the process of FRET.

If we assume that no significant quenching process is taking place then the lifetime is dominated by the electronic configuration of the molecular energy levels and the interaction of the molecule with the local environment. For every molecule there is an inherent fluorescence lifetime determined by the electronic structure of the molecule and the associated quantum mechanical rules. It is now possible to undertake complex calculations of a specific molecule and determine the expected fluorescence lifetime. Using these quantum mechanical models chemists are now able to modify molecules to alter their natural lifetime by adding or removing chemical groups and altering the electronic energy levels of the molecule. However, it is not possible to yet "dial up" the molecular structure of a fluorophore by putting in an excitation wavelength, emission wavelength and desired fluorescence lifetime, as all these parameters are linked.

This inherent, or natural, lifetime is altered by the environment surrounding the molecule. Before emission of the fluorescence photon some energy needs to be lost by the excited electron. The processes by which this happens can loosely be placed into two classes, physical changes and changes in the energy levels within the molecule. In the former group the excited electron loses some energy by transferring this to either other vibrational, or rotational, modes in the molecule, or to the local environment. Both of these routes for energy loss are affected by the viscosity of the environment and the temperature (and indeed these two effects are linked). If we have a molecule sitting in a very viscous solution it is clearly going to be harder for the molecule to vibrate, or for part of the molecule to rotate. Thus it will take longer for the excited electron to lose its energy by this route leading to an increase in lifetime. In a less "gloopy" world these movement modes are easier to form and for the electron to couple to these vibrational modes leading to a more rapid loss of energy and subsequent emission of the fluorescence. A much more detailed description of this process again can be found in textbooks on the subject (Lakowicz 1999).

The other class of loss mechanism is through the external environment slightly perturbing the electronic energy levels in the molecule. This alters their quantum mechanical overlap and the probability that an electron will lose energy. For example the ion concentration within a solution will place a local electric field around the molecule, marginally altering its energy levels. Many fluorophores thus have some sensitivity to pH, though in certain molecules, such as fluorescein, this effect is larger due to its structure and hence the interaction with the local electric fields. In a similar manner the lifetime can be affected by other chemical compounds, and also the polarity of the solution, surrounding the molecule.

The other major effect on the fluorescence lifetime of certain molecules is a combination of the physical and chemical environments. This is a prominent effect for molecules that bind to a specific molecular site; in biology this is frequently a protein. This binding can affect the molecule in two ways. It may alter its shape and hence its freedom to vibrate, changing the energy levels. Binding can also alter the electronic configuration of a molecule, or block access to ions present in the solution, leading to altered energy levels and a change in the lifetime.

Even with this simple description of some of the processes taking place within the fluorescence process it can be appreciated that observing the fluorescence lifetime can provide a wealth of information, but also that such data can be confusing. This means that, as with all experimental procedures, careful control experiments need to be undertaken before it can be certain exactly which parameter the lifetime is reporting from its local environment.

6.2 MEASUREMENT TECHNIQUES

In order to image using fluorescence lifetime we need to be able to know when a molecule is excited and how much later the fluorescence is subsequently emitted. As has been stated above, typically this process takes place on the nanosecond timescale and generally this means that high speed detectors and electronics are required along with light sources that can be switched rapidly. It should also be remembered that light has a maximum speed ($299,792,458$ ms^{-1}). Whilst in most microscopy applications this is so fast most events can be considered as instantaneous, this is not the case in lifetime measurements. In one nanosecond light travels around 300 mm and when one considers the optical beam path in most instruments is typically several hundreds of millimeters, this has to be taken into consideration in accurate lifetime measurements. Indeed, for a system with 100 ps resolution this distance reduces to only 30 mm. Generally in microscopy there are four methods used to image using lifetime as the contrast mechanism.

6.2.1 Time Correlated Single Photon Counting (TCSPC)

Time correlated single photon counting (TCSPC) is probably the most widely used lifetime measurement technique both for imaging applications and for the general determination of a fluorescence lifetime. It is currently the most accurate method of determining the lifetime but it may not be the fastest. The core methodology is, however, straightforward.

The basic technique is outlined in Figure 6.3. A short pulse of light (typically a few picoseconds or shorter from a laser) is directed onto the sample. As this pulse of light is emitted from the laser it starts a high-speed clock running. The pulse excites the fluorescence in the sample and this fluorescent light is collected and detected by a high-speed detector. The

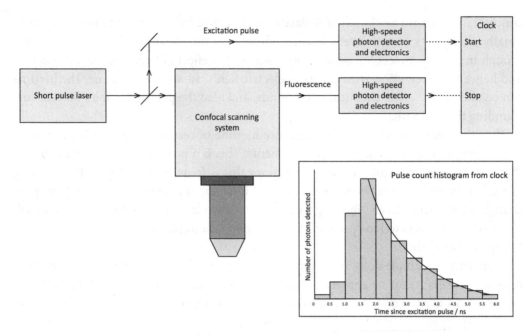

FIGURE 6.3   Diagrammatic representation of time correlated single photon counting.

moment the detector sees a photon it sends out an electrical pulse, which stops the clock and the time for this single photon event is recorded and placed in a histogram. A short while later another pulse is sent to excite fluorescence and the process repeated. Thus from each excitation pulse a single photon is detected and this enables a histogram of photon numbers received in a specific time window to be built up. After sufficient photons have been detected the histogram can be analyzed to provide the fluorescence lifetime as the emitted photons have a Poissonian distribution. In practice, the clock is frequently started by the arriving pixel, and the next excitation pulse used to stop the clock. As the light pulse frequency can be controlled with very high precision this does not affect the accuracy of the measurement.

The excitation pulse needs to be short, have a very rapid switch-off time and be able to operate at a high frequency (typically in the MHz region). Although it is possible to achieve this using LEDs (Rae et al. 2008) for nearly all imaging applications the source is a laser. Until the last few years these were typically mode-locked dye lasers requiring skilled operation and were expensive. However, two changes in technology have altered this situation. One was the arrival of non-linear microscopy (see Chapter 8), which requires an ultra-short pulse laser operating in the near infrared. Thus laser companies had a market for high cost and expensive laser systems and these became widely available. Although operating in the near infrared they could either be used directly to excite the fluorescence by multi-photon excitation or, using non-linear optical methods, frequency doubled to bring the wavelength down to the visible portion of the spectrum. More recently ultra-short pulse lasers have been produced which use non-linear effects in a special optical fibre to produce a white light spectrum of short-pulsed light. All these lasers typically have pulse lengths of 2–3 ps or less with repetition rates of several 10 s of MHz.

The other technical advance, and probably the most widely used in most imaging laboratories now, was the development of short wavelength visible laser diodes. Modulating a laser diode at high speed was well known but most devices operated in the red or near infrared portion of the spectrum. The advent of lasers based upon gallium nitride opened up wavelengths from the ultraviolet to green, the region of most commonly used fluorophores. Typical pulse lengths are around a few ps (thus generally longer than mode-locked lasers) but diode lasers have an advantage in that their repetition rate can easily be altered electronically to suit the fluorophore being measured.

The ability to alter the pulse repetition rate is potentially an important parameter. If the repetition rate is too high the next excitation pulse will arrive before the fluorescence from the first pulse has been emitted. This will clearly lead to incorrect lifetime measurements as the photon detected could have come from either excitation pulse. However, if the repetition rate is too low the time required to record a statistically sufficient number of photon counts in the histogram is significantly increased. As a rule of thumb the gap between the excitation pulses should be a minimum of three fluorescence lifetimes and ideally around five times the fluorescence lifetime.

The detector used for TCSPC clearly needs to be sensitive and have a fast response (the time of a photon arriving at the detector head and the electrical signal leaving the detector) and that transit time should be as consistent as possible. The traditional device for such measurements has been a specialized photomultiplier (see Chapter 2) though variations of the avalanche photodiode, and single photon avalanche photodiodes (SPADS) are now becoming more widely available. In the longer term, arrays of SPADS are likely to become the most common detector used for FLIM.

The output from the detector is then processed through several fast electronic circuits. Although different systems vary slightly, the first process is through a discrimination circuit, which only produces a pulse when the signal from the detector reaches a threshold value. The output is then turned into a single electronic pulse, which is the signal to either stop, or start, the clock. The time recorded on the clock is then used by the software to build up the histogram. The analysis and interpretation of these lifetime plots is described in Section 6.3.

In order to obtain an image using TCSPC a beam scanning system is required. The conventional FLIM instrument is built around a beam scanned confocal microscope with the standard laser being replaced by a suitable pulsed source along with a high-speed detector. The signals are then sent to the lifetime electronics board, which is synchronized with the scanning system. The beam is moved across the sample, but generally at a slower speed than in a conventional confocal microscope, as the system needs to build up a sufficient number of photon counts at each pixel position. The number of counts required for specific applications is discussed in Section 6.3.

6.2.2 Time Gating Electronics

TCSPC is the most accurate method of measuring a fluorescence lifetime, but it can be slow as each laser pulse only provides a single photon used to build up the image. This also means that it is inefficient in terms of its use of the fluorescence photons that have been

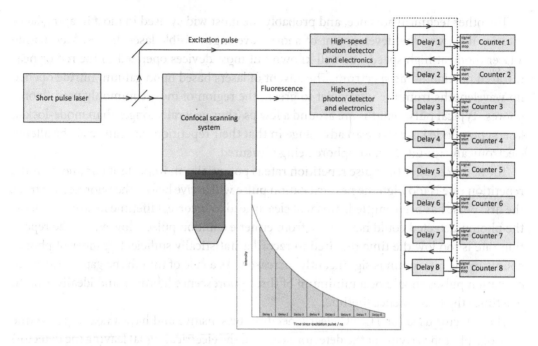

FIGURE 6.4   Diagrammatic representation of time bin gating.

generated. An alternative technique is to have fast electronics with a series of collecting bins and this method is illustrated in Figure 6.4. At the start of the excitation pulse the first time gate is opened and all signals detected by the fast photomultiplier are sent to the first electronic circuit. Here the electronics integrate all the signals received. After a short period, typically 1–2 ns for most common fluorophores, this first gate is closed and the signal is then directed to the second bin and integrator. This bin is then held open for a period of time before being closed and so on through the full array of time bins. The signals from the bins can then be collected and a histogram plotted (de Grauw and Gerritsen 2001).

The method does not have the absolute accuracy of TCSPC but even with a small number of bins (4–8) accurate lifetime images can be created. By increasing the bin number one can gain accuracy but the photons collected in each bin are fewer and thus it can take longer to build up a statistically significant lifetime image (Gerritsen et al. 2002). However, as will be discussed later, the absolute lifetime is frequently not the main reason for undertaking FLIM. The lifetime is being used as a contrast mechanism to separate out one feature from another, or alternatively one is looking for changes in the lifetime (as in the case of FRET) as an experiment progresses. Thus the binning method, although not widely adopted, has several advantages for FLIM.

6.2.3 Time Gated Camera

In the methods described above the imaging system is based around a confocal microscope and thus inherently involves beam scanning. This generally has a limitation on the speed, though with excellent spatial and temporal resolution. An alternative approach is to use an ultra-sensitive camera with a very fast shutter as the detector. The shutter can thus be kept open for a short time after the excitation light pulse, in the manner described above for a gated electron counting system. This single image will therefore have all the photons detected in a short time period. This can be repeated for a number of time periods to build up an image in a similar manner to the gating described above. Figure 6.5 illustrates a typical configuration of such a fast camera system.

The illumination source can either be a rapidly switched, high intensity LED or a beam expanded short pulsed laser. If a laser is used then the spatial coherence of the light has to be removed or the images will contain speckle. This is easily achieved using a spinning diffuser, though it does have the effect of scattering the light. The light is then sent into a conventional epi-fluorescence illumination system in a wide field microscope. The returned fluorescence light then comes back through the dichromatic filter and onto the camera. In the original method an image intensifier is fitted in front of the camera and the gain on this could be switched very rapidly to produce a time gate for light reaching the camera sensor. Thus a series of images can be taken at different time points from the excitation pulse and a lifetime image generated.

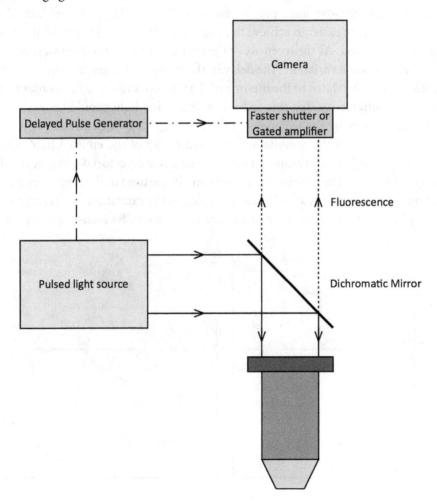

FIGURE 6.5 Optical configuration for wide field time resolved imaging.

In the last few years there has been a significant number of improvements in high-speed camera technology based upon the CMOS production process. This has led to significantly more sensitive sensors (scientific CMOS, or sCMOS, cameras) with high electronic amplification, while maintaining a low background noise level. As the electronics are all closely mounted to the sensor chip they can be adjusted to provide very rapid shutter speeds on the camera with time resolution down to 200 ps and timing jitter of around 10 ps enabling subnanosecond FLIM. These systems also tend to be lower in cost than conventional beam scanned systems with high-speed electronics and if a mass market opens up for such detectors then the camera cost could also fall significantly as the CMOS process is inherently designed for high volume, low cost production.

6.2.4 Phase Measurement

All the methods described so far have used a short pulse of light and a synchronized time gate or clock to determine the lifetime. An alternative method is to work in the frequency domain rather than direct timing. Although this method can be undertaken using a pulsed light source it has the advantage that it can be achieved using a sinusoidally changing intensity (which can be easier to achieve than single short pulses). Figure 6.6 illustrates the principle of the method. As the intensity of the excitation varies the fluorescence emission responds with the same variation. The delay in the emission is measured as the *phase shift* (ϕ) and this is directly related to the lifetime of the fluorophore present. In order to achieve an accurate measurement using this technique, excitation light should be modulated at a frequency that is given by the reciprocal of the lifetime ($1/\tau$). One can then utilize a high-speed detector to follow the modulation in the intensity of the emitted light, and using electronics (typically a lock-in amplifier or phase sensitive detector) determine the delay in the emission and hence the lifetime. A camera can also be used in the frequency domain by modulating the detector gain at the same frequency as the excitation source, again varying phase delay between the excitation and camera modulation. By adjusting the phase delay

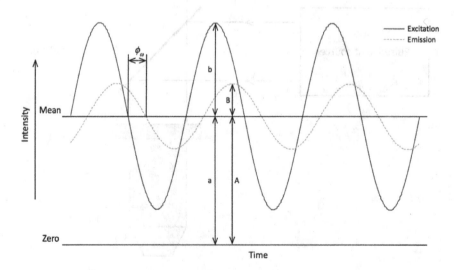

FIGURE 6.6   Phase modulation FLIM.

in this modulation relative to the excitation source until a maximum intensity image is obtained the lifetime can be determined.

For a system with a single exponential decay the finite time required for the sample to respond leads to a de-modulation of the emissions. This de-modulation is given by

$$m_\omega = \left(1+\omega^2\tau^2\right)^{-\frac{1}{2}}$$ (6.1)

and using Figure 6.6

$$m = \frac{B/A}{b/a}$$ (6.2)

As can be appreciated from the figure this value decreases from 1 to 0 as the modulation frequency is increased. The related value, the phase angle, is given by

$$\tan\left(\phi_\omega\right) = \omega\tau$$ (6.3)

The full derivation can be found in Lakowicz 1999, Section 5.11A.

Although not a widely used method the frequency domain measurement of lifetime can be useful in wide field imaging when one is looking for a change in the fluorescence lifetime rather than wanting to know the absolute value. Recovery of the lifetime from the data also becomes more complex when there are multiple lifetimes present within the sample, but by using modern computation techniques this is possible.

6.2.5 Slow Detector Method

In all of the methods described above the system has used a high-speed detector. However, it is possible to measure lifetimes in certain situations using a slow detector (Holton et al. 2009; Matthews et al. 2006). In principle even a webcam could be used in a FLIM system. The technique does require a short-pulse light source where one can alter the repetition frequency.

The basic concept is outlined in Figure 6.7. A very short pulse of light is used to excite the fluorescence within the sample. If the system is running at a low repetition frequency all of the fluorescence will have been released by the excited molecules before the next pulse arrives. However, at a high repetition frequency the next excitation pulse will arrive before all of the fluorescence has been emitted. Thus if one reduces the time between pulses further then less fluorescence will be emitted as there will be fewer molecules ready to absorb the excitation light and be re-excited. This means that as the repetition rate is increased there will be a further decrease in the integrated signal as shown in Figure 6.7b. In practice two images need to be recorded, each at a different frequency, and the ratio taken to provide an image in which the contrast is provided by the lifetime. While this does not provide a true lifetime image, with each pixel being assigned a specific fluorescence lifetime, it does provide a simple way of using the fluorescence decay as a contrast mechanism. For images captured with an exposure time of 20 ms one can then have lifetime images at video rates.

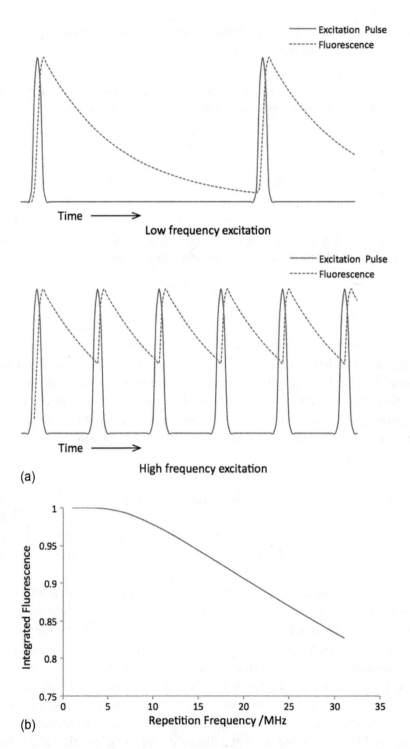

FIGURE 6.7   FLIM using slow detectors, a) fluorescence emission for short pulses at varying repetition frequencies, b) integrated fluorescence intensity as the repetition rate of excitation is adjusted.

There is one specific caveat to this mode of imaging. The fluorophores being imaged must not bleach during this repeated excitation. It assumes that the number of molecules available within the sample which can be excited does not change over time, and specifically between the two images used to take the ratio image. For many modern fluorophores this is a reasonable approximation and providing one has a suitable excitation source this can be a low-cost route to using lifetime as a contrast mechanism.

6.3 METHODS OF ANALYSIS

All commercial systems provide the relevant software for the construction and analysis of lifetime-based images. In such systems there is normally the opportunity to select the option to fit the data to a single or multiple exponential decay curve. The standard practice is to select a double exponential decay curve, though this can depend on the application. The image shown, however, will display the lifetime value at each pixel in the image normally shown through a colour encoded look-up table (Figure 6.8). This image should always be viewed alongside the intensity fluorescence image if possible, in particular if one wishes to link the lifetime to the physical structure being viewed.

Lifetime images frequently appear to have a lower spatial resolution than a more standard intensity image. Assuming that the images have the same number of pixels the actual spatial resolution is the same, but this can be obscured when the lifetime is used with a colour look-up table. This can be understood if one considers a standard image with 256 (8-bit) image depth (the intensity is divided into 256 intensity levels). If this is used to encode for intensity one has a large dynamic range. However, if the image is displaying lifetime from say 1 to 10 ns, each bit is around 4 ps in time and the image will appear noisy and blurred. As the temporal resolution is likely to be around 100 ps at best, a look-up table with a lower dynamic range should be used. This has the tendency to make the image appear to have lower spatial resolution, as a group of closely spaced pixels will all have the same colour (lifetime), whereas if intensity were used there would be greater detail present. However, lifetime and intensity images can be combined to link the lifetime features to the local spatial variation within the image.

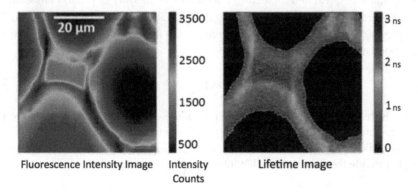

FIGURE 6.8 Fluorescence intensity and fluorescence image of endogenous fluorophores in Convallaria stems (Image courtesy of Dr Elvira Bohl, Edinburgh University).

As has been repeated several times in this chapter, and will be demonstrated in Section 6.4, one of the main roles of lifetime is a) to enhance the contrast when one has two fluorescent features that cannot be separated by wavelength, and b) to observe changes in the lifetime, rather than the absolute value. Thus the detailed analysis of FLIM images frequently requires the integration of spatial and temporal images. One also needs to be careful when considering changes in the lifetime to ensure that the change being seen is significant. As discussed below the software available for the analysis of lifetime data enables a number of parameters (typically how many fluorescence lifetime or exponential terms there may be) to be fitted. Care has to be taken that when looking at changes, the error within the lifetime is considered very carefully, and that the fitting parameter used (number of exponential terms) can be justified. This really means applying good scientific and statistical judgment to the data and not just selecting fitting parameters until the most interesting answer emerges, which has been known to happen.

What does the fluorescence lifetime τ that has been used throughout this chapter actually represent? If we have a fluorescent sample, which is excited by an infinitely short (delta function) pulse of light, this will provide an initial population of excited molecules of n_0. This population of excited molecules will decay by a rate $\Gamma + k_{nr}$ given by the equation

$$\frac{dn(t)}{dt} = -(\Gamma + k_{nr})n(t) \qquad (6.4)$$

where $n(t)$ is the number of molecules excited at time t, Γ is the rate of fluorescence emission and k_{nr} is the non-radiative decay rate (loss of energy by routes other than the emission of a photon). The fluorescence emission is a random event with each excited molecule having the same probability of emitting a photon at a given time point. This leads to an exponential loss of the excited state population with the population at any point in time after excitation given by

$$n(t) = n_0 \exp\left(-t/_\tau\right) \qquad (6.5)$$

where n_0 is the initial number of molecules excited.

In any of the methods used to determine the fluorescence lifetime we do not directly observe the number of excited molecules but rather the intensity of photon emission, which is clearly proportional to the number of excited molecules at any one time $n(t)$. By using this fact and integrating equation 6.4 we can determine the intensity of light being emitted at any particular point after excitation given by

$$I(t) = I_0 \exp\left(t/_\tau\right) \qquad (6.6)$$

where I_0 is the intensity at the time of the end of the excitation pulse. The lifetime τ is the inverse of the total decay rate containing the radiative and non-radiative terms. It is this exponential loss of fluorescence intensity that is then plotted based upon the detected photon numbers to provide the fluorescence lifetime image.

The exponential decay of detected fluorescence contains all the routes by which an excited fluorophore can lose its energy. One therefore has to decide when fitting the data if a single or multi-exponential decay should be used. In general, except in the case of FRET or other quenching processes, the non-radiative decay term is small and is thus frequently ignored in data analysis. However, the most common analysis method is a bi-exponential fit to the data to cover the two standard methods of energy loss. For example, this is particularly important if the lifetime is being used to determine changes in the viscosity, which affects the non-radiative term.

It is also possible to include higher order exponential terms if one has a reason to believe that the decay process will be affected by multiple factors. Clearly the greater the number of exponential terms used the greater the number of photons are required to obtain a statistically good fit to the data. This means that a longer pixel dwell time may be required at each point in a beam scanned image leading to greater imaging times. In addition there are alternative methods, which can be used to analyze the data provided including *phasor analysis*. Readers interested in alternative routes are advised to look at more advanced texts on lifetime measurements (Becker 2005).

In summary, in any imaging application using FLIM one needs to consider exactly what information is required to answer the question being posed. In all FLIM methods one is calculating an average lifetime, or lifetimes, and thus careful analysis of the statistics is required before definitive conclusions can be drawn. If one is using the lifetime purely as a contrast mechanism, or looking simply for a change in a lifetime, then the imaging can be faster as the statistical fit need not be as accurate.

6.4 EXPERIMENTAL CONSIDERATIONS AND GUIDELINES FOR USE

With the basic principles and technical consideration behind FLIM having been explained we now need to consider the most appropriate applications in which to use this technique and some of the practical implications of the method in terms of experimental design and sample preparation. As with all imaging techniques the fundamental axiom is to use the simplest method which will minimize the perturbation to the sample while providing data which is of sufficient quality to answer the questions of interest.

Before considering the positives of FLIM the limitations should be appreciated. Generally the technique is not going to provide high-speed images, even with a wide field camera-based system. Frame rates above one to two frames per second are unusual. This is mainly because of the requirement to have sufficient photons counted to provide a good statistical basis upon which the lifetime is calculated. The measured lifetime is affected by multiple parameters and thus assigning a specific effect to a lifetime change requires complex controls and consideration of all the experimental parameters. In addition, although not a direct technical reason for not selecting the method, the equipment is not simple to operate and at the present time the instruments are costly.

However, using lifetime as a parameter in a microscopy system does enable a wide range of experimental procedures to be undertaken, that would be harder, or in some cases impossible, using just intensity as the contrast mechanism. It is also worthwhile noting that where FLIM is not used as the standard imaging method in an experimental protocol it can nevertheless provide significant additional information to help explain a process that has been observed using another method. There are probably three major roles to consider for FLIM and each of these is outlined below, illustrating the specific features of the FLIM method and why it might be selected.

6.4.1 FLIM for Enhancing Contrast

Although there is a very wide range of fluorescent markers now available there are occasions when indigenous fluorescence within the sample is excited and emitting at wavelengths very close to the fluorophores that one wishes to observe. This happens frequently in plants where the natural light absorbing, and subsequent emitting, properties of chloroplasts can make fluorescence imaging difficult. The same is also true in certain *in vivo* imaging applications where collagen or elastin, for example, may be present causing a background fluorescence which cannot be removed through more conventional wavelength separation techniques. In these applications the lifetime is being used as a contrast mechanism rather than actually interpreting the lifetime as a parameter in the experiment.

An example of the use of this method is imaging within the live lung where elastin, which has very broad absorption and emission, can be present as well as a fluorescent marker introduced to detect a specific disease. The elastin signal is useful as it provides a background image to help navigation around the lung and therefore one does not want to remove the signal completely. The lifetime of elastin is around 2–3 ns while that of the fluorophores being introduced can be 6–8 ns with the exact value depending on the specific fluorophore selected. Thus one can separate the two sources of fluorescence by using a simple binning method in which all the early photons (those before 3 ns) are placed in bin one and assumed to come from the elastin, and the remaining photons in a second bin, with a window from 6–8 ns, assumed to have come from the introduced fluorophore. One now has a high contrast image that can be subsequently analyzed with the tissue structure clearly separated from the introduced fluorescent label. A similar technique can be used for removing, or separating, background fluorescence in other systems.

This simple binning operation can also be used to remove fluorescence from Raman-based imaging methods (see Chapter 10). Here the low level Raman signal appears "instantaneously" as it is a scattering process and the fluorescence will appear after a few nanoseconds. Thus using the same technique the fluorescence can be gated out of the image. With an increasing interest in Raman-based analysis methods it is anticipated that Raman microscopy will also grow and thus the time gated contrast method is likely to become more widely used.

6.4.2 FLIM for FRET and Observing Changes in Lifetime

A major application for FLIM is to observe changes in the fluorescence lifetime. Here the absolute lifetime is not of interest but it is local changes that are important. These changes may represent alterations in protein structure, local environmental conditions or chemical activity. The data analysis may be straightforward in this situation using the binning method described above and one simply monitors changes in the detected lifetime.

One of the major current uses for FLIM in this mode of operation is to observe changes in fluorescence lifetime caused by the FRET processes. This was mentioned

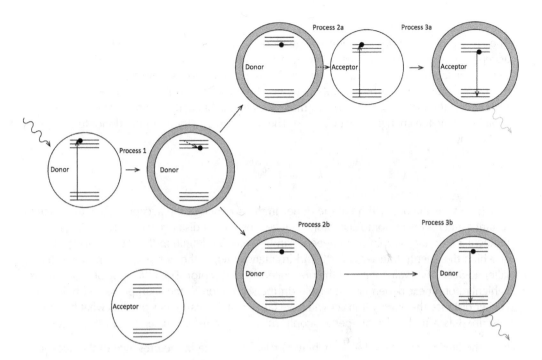

FIGURE 6.9 Representation of the FRET processes.

briefly early in the chapter but the details are now presented. The basic concept is outlined in Figure 6.9 and is based upon the quenching of fluorescence by the presence of a second local molecule.

Initially the first, *donor* molecule absorbs light and is excited to a fluorescent state (Process 1 in Figure 6.9). However, a second molecule, the *acceptor*, is close by and the excited state energy is transferred (Process 2a). The acceptor molecule, now in an excited state, then emits fluorescence at its emission wavelength, which will be longer than the donor (Process 3a).

The acceptor molecule has an absorption profile that is spectrally well matched to the donor's emission profile. For example one FRET pair using common genetically encoded fluorophores is cyan and green fluorescent proteins, and one using more conventional molecular pairs is fluorescein and tetramethylrhodamine (TMRE). In the FRET process no actual photon emission takes place but the energy of the excited donor molecule is transferred to the acceptor through electromagnetic interactions.

Not all excited molecules will transfer their energy, and the level of energy transferred depends on inverse spatial separation of the two molecules raised to the sixth power. If the acceptor molecule is too far away (Process 2b) then the donor molecule will emit fluorescence in the normal manner (Process 3b). The resonant energy exchange is very much faster than the fluorescence process and thus the fluorescence lifetime of the donor molecule is shortened from its natural "isolated" state. Thus by observing the lifetime the level of FRET can be determined and the spatial separation of the two molecules inferred.

The FRET process is due to the long-range dipole-dipole interaction between the donor and acceptor molecules. This process is much faster than fluorescence emission and depends on two major factors, the spectral overlap of the donor emission spectrum with the acceptor excitation spectrum and the separate between the donor and emission. In most FRET applications it is the distance between two molecules that is being determined, or in the case of protein folding the change in that distance. The rate of energy transfer from donor to acceptor is given by

$$k_T = \frac{1}{\tau_D}\left(\frac{R_0}{r}\right)^6$$

where τ_D is the decay time of the donor in the absence of the acceptor, R_0 is the Förster distance and r is the donor to acceptor distance. The Förster distance varies depending on the molecular pair and is typically between 2 and 10 nm. By definition this distance is the point at which the energy transfer is 50% and in intensity terms the donor emission would have fallen to 50% of its value without the presence of the acceptor. The efficiency of this process is highly non-linear depending inversely on the separation to the sixth power. At twice the Förster distance the energy transfer will have fallen to 1.56% and thus one has what has been described as a molecular or spectroscopic ruler. The transfer efficiency can be calculated from the lifetime using $E = 1 - \dfrac{\tau_{DA}}{\tau_D}$ where τ_{DA} is the lifetime in the presence of the acceptor and τ_D the lifetime of the donor in isolation.

In this equation relating the efficiency to the lifetime we have assumed that the decay of the donor in isolation is a single exponential function (which we know is not strictly true), but if one uses the average lifetime the approximation is generally robust. A full explanation of the FRET process is beyond the scope of a book on microscopy and the reader is referred to more detailed texts on the FRET process though a good background start can be found in chapters 13 to 15 of Lakowicz (1999).

The alternative to using FLIM for FRET-based imaging is an intensity ratio change between the two fluorophores' emissions. However, this can be inaccurate as the change in intensity may only be a few percent, and care has to be taken to be sure about changes in the excitation intensity and also the risk of photobleaching of one of the fluorophores.

6.4.3 FLIM for Absolute Lifetime Measurement

The chapter so far has focused on the use of the fluorescence lifetime as a method of increasing contrast within images or for FRET-based experiments. These are probably the most frequent applications of FLIM in current use. However, there are applications of FLIM in which the exact lifetime is measured and this can provide significant information on the local environment within a sample, and specifically inside a cell.

It has been stressed throughout that the fluorophore needs to lose a certain quantity of energy before the fluorescence emission can take place and that this is altered by the local conditions; thus with careful calibration certain properties of the local environment can be measured. For most fluorophores the major environmental change that is monitored is the viscosity of the solution, though as mentioned earlier localized ion concentration will also affect some fluorophores. It is not the aim here to provide details on the process but to make the reader aware that by measuring the lifetime accurately, and carefully calibrating the lifetime of a fluorophore against the variable of interest, an estimate can be made of the local environment. In making these measurements one can also look at the change in polarization of the emitted light compared to the excitation pulse in a method known as polarization anisotropy. This clearly reduces the number of photons being detected and thus is not normally used alongside FLIM as it can significantly increase the time required for image acquisition.

6.4.4 Practical Considerations

As has been described above there are a number of practical situations in which FLIM is clearly the method of choice to reduce the background signal from overlapping fluorophores, monitor the FRET process and determine something about the local environment. All FLIM imaging systems come with comprehensive instructions and training. However, it is worth listing a few general experimental factors that should be considered before, and whilst, undertaking FLIM though many have been mentioned in the preceding text.

1. *Pulse repetition rate*: This should be rapid enough to obtain images at a reasonable speed but crucially should not be so high that the next excitation pulse is received before the entire emission from the previous pulse has taken place. A minimum time of three fluorescence lifetimes should be allowed between pulses and ideally five.

2. *Excitation intensity*: As with all methods the aim is to keep the excitation power as low as possible to obtain a good enough image. However, in the case of FLIM it is important not to have too high an excitation pulse intensity as this can lead to excessively high count rates. It is possible that even with the high-speed electronics discussed earlier the fluorescence photons arrive so quickly that the detector will not be ready to detect the next photon. This means that the early part of the histogram will reach very high numbers quickly leading to a distorted decay profile. Thus each FLIM system will have a suggested maximum count rate to prevent the so-called "pulse pile up" from affecting the lifetime measurement. Typically the count rate should be between 1 to 10% of the excitation rate.

3. *How many photons to detect*: This is clearly dependent on the accuracy required in the FLIM image but using TCSPC without binning one typically requires around 10,000 photons to be detected for an accurate lifetime measurement. Using binning methods, and if one is only looking for contrast enhancement from the image, 1,000 photons can be sufficient.

4. *Instrument function*: As mentioned earlier, in 1 ns light travels around 300 mm and for fluorescence microscopes with optical paths of several hundred millimeters the distance that the light pulses travel both for excitation and detection needs to be allowed for in any measurements. A simple measurement for this, and one that should be checked occasionally, is to use light reflected from a glass coverslip to calibrate the system. Reflection is an instantaneous process and thus can be used to

determine how long the light takes to reach the detector, and for the electronics to process the data. This then sets the zero position on the lifetime histogram. The excitation pulse will also typically be around 1–2 ps or less and thus is a single time spike for the electronics and hence the accuracy of measurement of the system can also be determined.

Fluorescence lifetime imaging is thereby a method that can provide significant information beyond a simple intensity–based image. The downside is that the equipment is complex, costly and the imaging inherently slower than that for other more conventional forms of microscopy.

REFERENCES

Becker, W. 2005. *Advanced Time-Correlated Single Photon Counting Techniques.* Springer Berlin Heidelberg.

Gerritsen, H. C., M. A. Asselbergs, A. V. Agronskaia and W. G. Van Sark. 2002. "Fluorescence Lifetime Imaging in Scanning Microscopes: Acquisition Speed, Photon Economy and Lifetime Resolution". *Journal of Microscopy* 206(3): 218–24.

de Grauw, C. J. and H. C. Gerritsen. 2001. "Multiple Time-Gate Module for Fluorescence Lifetime Imaging". *Applied Spectroscopy* 55(6): 670–8.

Herschel, J. F. W.. 1845. "On a Case of Superficial Colour Presented by a Homogenous Liquid Internally Colourless". *Philosophical Transactions of the Royal Society of London. Series B, Biological Sciences* 135: 143–5.

Holton, M. D., O. R., Silvestre, R. J. Errington, P. J. Smith, D. R. Matthews, P. Rees and H. D. Summers. 2009. "Stroboscopic Fluorescence Lifetime Imaging". *Optics Express* 17(7): 5205–16. www.ncbi.nlm.nih.gov/pubmed/19333284.

Lakowicz, J. R. 1999. *Principles of Fluorescence Spectroscopy.* 2nd edn. Kluewer.

Matthews, D. R., H. D. Summers, K. Njoh, R. J. Errington, P. J. Smith, P. Barber, S. Ameer-Beg and B. Vojnovis. 2006. "Technique for Measurement of Fluorescent Lifetime by Use of Stroboscopic Excitation and Continuous-Wave Detection". *Applied Optics* 45(9): 2115–23.

Rae, B. R., C. Griffin, J. McKendry, J. M. Girkin, H. X. Zhang, E. Gu, D. Renshaw, et al. 2008. "CMOS Driven Micro-Pixel LEDs Integrated with Single Photon Avalanche Diodes for Time Resolved Fluorescence Measurements". *Journal of Physics D: Applied Physics* 41(9): 94011. http://stacks.iop.org/0022-3727/41/i=9/a=094011?key=crossref.9b34cef525d6f0172059d12104636b93.

CHAPTER 7

Light Sheet or Selective Plane Microscopy

7.1 INTRODUCTION

Since its reinvention in around 2004 light sheet microscopy has rapidly established itself as the method of choice for imaging live, essentially transparent, samples. The very rapid acceptance of the method is linked not just with improvements in technology, in particular cameras, but also with the increasing desire for extended imaging of *in vivo* models such as the zebrafish, *Drosophila* and germinating plants. Light sheet microscopy enables high-speed, three-dimensional imaging for extended periods of time, and when this is linked with biological techniques, such as the use of fluorescent proteins, dynamic processes can be studied with both high temporal and spatial resolution. The initial application for the method was in developmental biology but the use has now spread across many fields and includes neuronal studies, immunology (both *in vivo* and in living, dissected organs), tissue repair, cardiology, ophthalmology and plant science. As will be seen throughout this chapter the growth has been stimulated by the OpenSPIM initiative (Pitrone et al. 2013). The fundamental optical concepts of the technique are simple, requiring just basic optics and a camera, therefore high-performing instruments can be built easily in a user's laboratory for a reasonable price. The full system design, building instructions and control software have been made available in an open source manner. This route to owning a system has also stimulated numerous interactions between life scientists (who want a system but perhaps lack the confidence to build their own) with optical physicists who then discover the stimulating environment of interdisciplinary research.

Before the basic concept of light sheet microscopy is described, the nomenclature should be considered briefly. The original method, using a sheet of light rather than the more conventional patch or single spot, was known as "ultra-microscopy". When this method was rediscovered the initial publications used the term "light sheet" before selective plane illumination microscopy was introduced, which was also described as single plane illumination microscopy. As it fits both descriptions the acronym SPIM has now become adopted to describe numerous variations of light sheet imaging, and thus this is the name that will be used throughout this chapter to describe the core imaging technique.

The basic SPIM concept is outlined in Figure 7.1. A thin sheet of light is directed into the sample that in most modern practical applications excites fluorescence in a single plane. The fluorescence, however, is emitted in all directions and hence an imaging lens can be placed at right angles to the light sheet to collect the light emitted in that direction. This orthogonal imaging system thus collects all of the light from the single, or selected, plane that has been illuminated. The collection optics must be focused in the same plane as the illumination and therefore, to produce a three-dimensional image, the easiest practical solution is then to scan the sample through this common focal plane recording an image for each optical section.

This configuration has several significant advantages over more conventional optical sectioning imaging techniques such as confocal microscopy. The excitation light only illuminates the section of the sample that one wishes to image, whereas in confocal imaging a cone of the sample is illuminated and only light from the focus detected (see Chapter 5). The SPIM configuration is thus more light efficient in that all the excitation, and hence emission, in the field of view is used. Fluorophores are only excited when they are going to be imaged, reducing photobleaching and other unwanted phototoxic events. As an entire optical section is captured in a single exposure (compared to a confocal system in which the beam is scanned) the imaging is inherently faster and a high-speed camera can be used to record the images. The method has thus minimized the light exposure on the sample and developed the ability to image intact samples at high speed, making it an ideal method, for example, in developmental biology.

However, there are some limitations. The sample needs to be basically transparent and not highly scattering as the illumination light requires a clear path to the volume to be

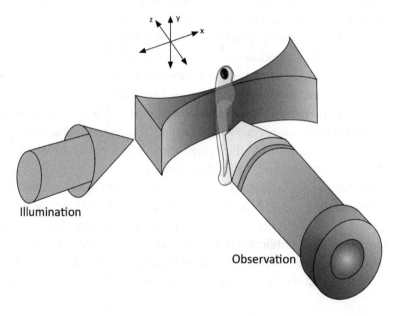

FIGURE 7.1 Basic concept of light sheet of SPIM imaging.

imaged. If there is a region which absorbs light before the imaging plane then dark streaks will appear on the image (see Section 7.5 for figures illustrating SPIM limitations and faults). If incoming light is scattered it can lead to out of plane fluorescence and hence loss of contrast in the images. Emitted fluorescence from the plane of interest can also be scattered, or aberrated, on the way to the detector. The optical sectioning resolution is determined by the thickness of the light sheet which, as explained below, is limited and does not produce as fine an optical section as can be achieved with a true confocal system. There are also a few physical considerations; the requirement for two objective lenses to be mounted orthogonally and focused to the same point in space limits which lenses can be selected. This is due to working distance considerations, the physical proximity of the lenses and the need to generally move the sample through the imaging volume. However, the speed and minimal level of perturbation which enables long-term imaging make SPIM a very attractive method and there are ways around these limitations. In summary if you want to image intact, frequently *in vivo* samples, at high speed and/or for extended periods of time, SPIM provides an excellent technique to consider.

This chapter will now look briefly at the history of the development before explaining the basic optical principles together with the design and application of a practical system and the basic configurations. This will be followed by practical details on sample mounting, a very important consideration in light sheet microscopy, and some associated techniques for long period *in vivo* imaging. Data processing and more advanced imaging methods will then be discussed before looking at some of the limitations of the method. There is an extensive and rapidly growing literature in this field and once readers have appreciated the major concepts of light sheet imaging they are advised to search for previous work in their specific application.

7.2 BRIEF HISTORY OF LIGHT SHEET MICROSCOPY

The first reported microscopy system which used a sheet of light was in 1902 when a German Chemist, Richard Zsigmondy, and an optical physicist working at Zeiss, Henry Siedentopf, developed what was known as the "ultra-microscope" (Siedentopf and Zsigmondy 1902). This was designed, patented and built to help understand colloidal materials, in particular gold, and the results from this work earned Zsigmondy the Nobel Prize for Chemistry in 1925. The philosophy of the system was that for high quality images one requires high contrast, thus the aim was to remove out of focus light so that all that was observed was light scattered towards to observer. Their system contained a method to produce a light sheet onto the sample and the light that was scattered, in so-called Tyndall cones, was observed through an objective mounted orthogonally to the sheet of light. This system worked well and was also used by others to study Brownian motion. However, this was perhaps an example of an invention ahead of its time as the technology to exploit the method was not available. Photography at this stage was limited to photographic film and even moving images were captured by hand-cranked cameras and required exceptionally high light levels. The idea of an electronic camera was still around thirty years away and thus the method was never widely adopted.

It was around ninety years later when the method was "rediscovered" and the light sheet was used to reduce the excitation volume within a sample (Stelzer and Lindek 1994;

Voie, Burns and Spelman 1993). Technology was now readily available to exploit the break-through with the availability of personal computers and CCD cameras. However, the real publication that caused the wide-scale uptake of the method was in 2004 where it was used *in vivo* (Huisken et al. 2004). This technical development also came at a time when "labelling" via fluorescent proteins had become standard practice and there was a growing interest in the use of models such as the zebrafish for developmental studies. These samples utilize all of the positive features for SPIM described above and when allied with further technical developments, and the release of the OpenSPIM documentation, the technique saw very rapid growth. As the basic configuration is simple, systems can be constructed with off the shelf components and hence the entry to the technique was lower than for other forms of optical sectioning microscopy. The flexibility offered by the configuration, that of having the excitation and detection in different optical paths, has led to a wide array of different forms of the instrument and the ability for researchers to optimize the core method for their specific application.

7.3 OPTICAL PRINCIPLES

The key to SPIM is the formation of a thin, uniform sheet of light as illustrated in Figure 7.1. However, this does mean that compromises need to be made as the thinner the sheet, in general, the smaller the length over which it is thin. In the most commonly used configuration the light sheet is actually achieved through what might be termed "abuse" of the focusing power of a microscope objective. A conceptual optical diagram is shown in Figure 7.2a. The light, typically from a laser due to the requirements for spectral brightness, is initially expanded and then passed through a cylindrical lens. This produces a line of light at the focus of the lens and this line of light is then re-imaged onto the back aperture of the illuminating objective lens (Figure 7.2a inset).

Figure 7.2b separates the two optical planes of the cylindrical lens. In one plane the back aperture of the objective lens is not filled, meaning there is a very low effective numerical aperture leading to minimal focusing in this plane. Thus the height of the beam at the back of the objective is hardly altered as it goes through the lens, setting the vertical extent of the light sheet. In the other axis the objective lens will use its full optical power (numerical aperture) to produce a focus as the back aperture is filled. This focusing determines the thickness of the sheet and is set predominantly by the numerical aperture of the illumination objective. To produce a thin light sheet one therefore needs an objective lens with a high numerical aperture. However, the higher the numerical aperture the shorter the depth of field and thus the light sheet will only be thin over a small part of the imaging objective's field of view. Without using exotic beam profiles (see Section 7.7) one therefore has a compromise over the field of view and the optical section thickness. The imaging lens numerical aperture also affects the actual effective thickness of the optical section observed as this sets the efficiency of collection of the light in the axial imaging direction (Engelbrecht and Stelzer 2006). In a typical SPIM system the imaging objective has a numerical aperture of around 0.8 which when combined with the light sheet in a typical system produces a field of view of around 150 µ and an optical section thickness of 1–2 µ.

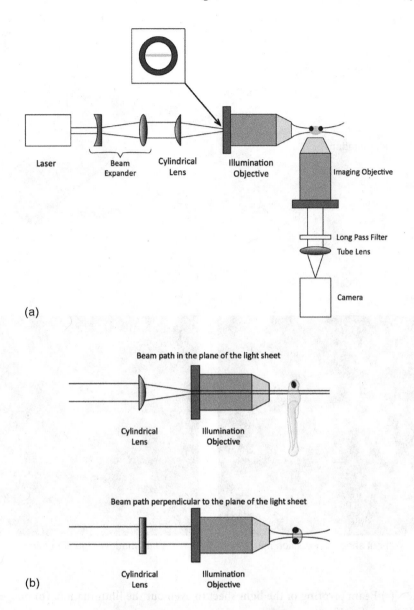

FIGURE 7.2 Conventional configuration of an SPIM illumination system: a) shows the basic configuration with the inset illustrating the shape of the beam at the back aperture of the illumination objective, b) illustrates the different focusing properties of the illumination beam caused by the cylindrical lens.

One of the limitations of SPIM, as mentioned above, is the loss of excitation light before reaching the imaging volume due to absorption. This leads to dark streaks on the image due to a lack of excitation. If the absorption is caused by a small absorbing volume, then rapidly scanning the beam, pivoting around the imaging volume of the system, can average out the illumination. Figure 7.3 illustrates the principle in which a rapidly scanning mirror (typically a resonant scanner running at around 4–8 kHz) is re-imaged so that its movement pivots the beam in the focal volume in the vertical direction. The beam is

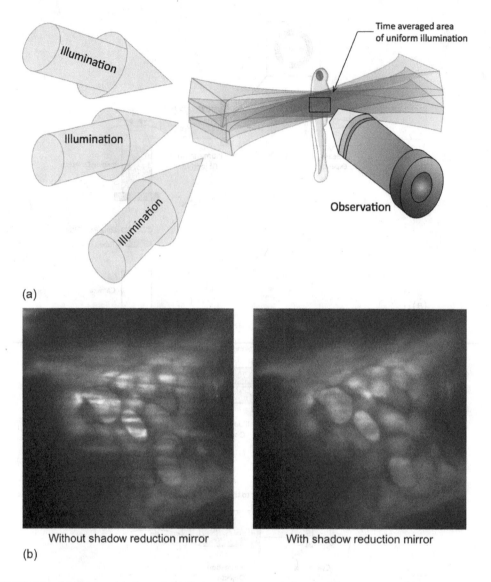

(a)

Without shadow reduction mirror   With shadow reduction mirror

(b)

FIGURE 7.3   (a) Beam pivoting of the light sheet to even out the illumination, (b) image example showing the type of artifact removed.

scanned much faster than the camera exposure time leading to illumination on the sample coming from multiple angles and averaging to provide even illumination by navigating around the areas of absorption present for a non-scanning beam.

The orthogonal imaging arm in a SPIM system determines the lateral resolution of the system and this is determined entirely by the imaging lens numerical aperture. Using the typical 0.8 NA lens mentioned above provides a lateral resolution of around 0.4 µm for a green fluorescent protein labelled sample with a peak fluorescence emission at around 560 nm. Using a series of dichromatic filters it is then possible to develop an imaging system capable of multi-wavelength imaging (Figure 7.4). As in all modern microscopes that use a camera, the pixel size and magnification of the system also has a role to play in the

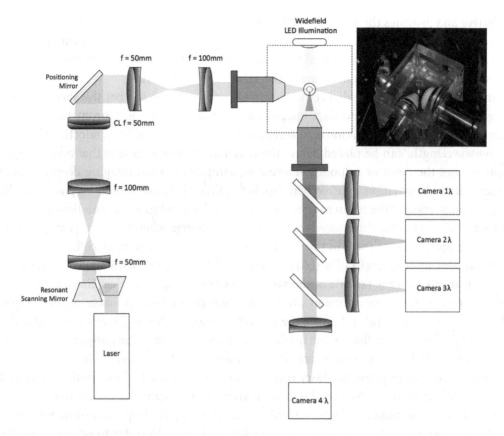

FIGURE 7.4 Optical configuration of a practical SPIM microscope used in the author's laboratory with the inset showing the lens mounting using O-rings into the sample chamber. The sample is mounted from above in this case using FEP tubing.

final resolution (see Chapter 2). It should also be noted that the magnification of a specific microscope objective is only valid when used with the correct "tube" lens before the camera and that different manufacturers design their objectives to work with different focal length tube lenses.

7.4 PRACTICAL SYSTEMS

7.4.1 Optical Details

Having now explained the principles behind SPIM, how is this actually achieved in practice? Figure 7.4 shows the components used in a typical SPIM system and is based upon published designs (Greger, Swoger and Stelzer 2007; Pitrone et al. 2013) and instructions are available on the OpenSPIM website for the construction of a fully functioning system. In the systems built and in current use in the author's laboratory (Huisken and Stainier 2007; Taylor, Girkin and Love 2012) light from a laser is directed onto a resonant scanning mirror operating at around 8 kHz. This mirror is conjugated to the image plane of the illumination objective. A second mirror is introduced into the system at the focal point of the cylindrical lens and this mirror is conjugated to the pupil plane of the illumination objective and provides the ability to adjust the "z" position of the light sheet so that it can be adjusted to be in the same focal plane as the imaging objective. The lenses are placed to ensure that the optics are conjugated to the correct focal planes and provide the magnification required to fill the back aperture of the microscope objective.

In the imaging path the dichromatic mirrors are arranged so that the shorter wavelengths are reflected first and then imaged onto the cameras using achromatic lenses (to minimize achromatic aberrations). In some cases an extra spectral filter, rejecting the excitation wavelength, can be placed before the lens and camera to ensure that no excitation light reaches the detector in poorly fluorescing samples or when imaging deeply to help reject scattered excitation light. When this is required the filter should be placed in the collimated portion of the beam as spectral filters are designed to work with light hitting at normal incidence since their wavelength properties change when the light is at an angle. The other optical item shown in the diagram, which is not present in all SPIM systems, is an arrangement of several LEDs (blue, green and IR) along the same optical axis as the imaging arm but on the far side of the sample from the imaging optics. These are used to create a conventional transmission (in the infrared) or fluorescent image of the sample. This feature is very useful in finding the correct position within the sample, in particular when only a few cells are fluorescently labelled in a specific area of the sample. The infrared LED is also used in the synchronization method described in Section 7.5.4.

In the most basic system the laser illumination can be switched on continuously and the imaging cameras run to record images in an open manner using the control software provided with the cameras. However, this is not the preferred option as one wishes to expose the sample to the minimum level of light possible. In order to achieve this the laser light can be activated just before the camera shutter is opened and synchronized to the camera shutter and thus no photons are wasted. Using the various all-solid-state laser sources that are now available this switching of the excitation light is routinely achieved using simple digital logic controlled either by the computer or a separate trigger circuit. It should be noted at this point that when operating in this "switching mode" a typical SPIM system will record images at around 5–15 Hz. The flashing laser light at this frequency, even when carefully shielded for laser safety reasons, can be seen in a darkened room and can induce nausea in some people, thus full shielding is advised where possible. Once the limits of the image stack have been determined manually then the full 3D data set can be recorded by running the sample through the excitation light sheet using the motorized drive along the imaging optical axis. A standard imaging protocol is given in Section 7.5.

The main variation on this basic method is the use of the resonant mirror to tilt the light sheet in a vertical plane rotating about the centre of the imaging field of view. Operation of this mirror helps to remove the absorption induced "streaks" that can be seen in certain SPIM images (Figure 7.3). The main downside of this procedure is the constant high-pitched noise from the resonant mirror but the benefit in image quality in many samples is worth the effort. In typical laboratory use the mirror is not active during the alignment and initial search for the best imaging area but is then activated for the actual imaging process.

7.4.2 Basic Alignment

As with all optical microscopes the quality of the images is clearly influenced by the alignment of the optics and a SPIM system introduces new challenges compared to a more conventional epi-illumination microscope, even those of a beam scanned system. The main challenge is to obtain a high quality, uniform light sheet that is in the same focal plane as the imaging objective, throughout the field of view. The following provides one protocol that can help with the initial daily alignment checks and assumes that the basic optical system with the various optical planes being re-imaged has been correctly set up.

1. A miniature "pentaprism" with an aperture of around 2–5 mm can be mounted on the sample holder and inserted into most SPIM systems. A pentaprism (used in the view finders of SLR cameras) turns light through exactly 90° and should be adjusted such that the excitation beam passes into the entrance aperture.

2. The detection optics should be adjusted so that light reaches one of the cameras. This may entail removing the dichromatic mirrors, but will depend on the exact wavelength of the mirrors and the laser wavelength.

3. The excitation beam can then be adjusted for position to produce a uniform illumination on the camera. During this process it is often useful to change the look-up table (LUT) that the camera software is using to display the image such that intensity is converted to colour. Exactly which optical element should be adjusted to correct a specific non-uniformity in the light sheet is learnt by experience and does vary from system to system.

4. The pentaprism should then be removed and replaced by a low concentration (a dilution of around 1:40,000) of fluorescent beads mounted in low melting point agar. If the system is a multiple wavelength microscope then a range of different fluorescent beads should be used. From experience beads at around 2 μm are easy enough to find whilst being small enough to provide a reliable test of resolution. If removed earlier the dichromatic mirrors should now be replaced.

5. The beads should then be viewed (ideally using the same laser power and camera parameters) to monitor long-term drift and the focus of the system adjusted (as the pentaprism is not really affected by the focus of the imaging optics). In most SPIM systems this is achieved by adjusting the lenses before the cameras as the imaging objective is frequently fixed into the sample tank with "O-rings". In multiple wavelength systems several of the beads should appear in more than one camera and by overlaying the images the different cameras can be accurately registered. It may also be necessary to rotate the cameras slightly to achieve good alignment.

6. A sample z-scan should then be undertaken and the data set examined quickly to ensure that the resolution is suitable in all dimensions (see Chapter 3 for details). Such images can also be used for spatial calibration purposes later.

7. The system is now ready for use. Further potential "imaging defects" and corrections are discussed in Section 7.5.3.

7.4.3 Basic Variations in SPIM

There are now numerous variations of SPIM systems (Figure 7.5), but as with most forms of optical microscopy there is one decision that does need to be made early on, that of conventional or inverted SPIM (Wu et al. 2011). This is largely driven by the sample that one wishes to image. The standard configuration for SPIM, and that used in most diagrams, is with the light sheet coming in parallel with the table to produce a vertical sheet that is then viewed with a camera mounted parallel with the table. Optically this is the simplest system to build (components are easy to mount to an optical table or breadboard) but the sample then has to be lowered into the excitation/imaging volume from above. Typically it is suspended in agar (see Section 7.5.1) as one is fighting the effects of gravity wanting to pull the sample down. In an inverted SPIM system (iSPIM) the illumination and detection optics are mounted vertically, each around 45° to the table, as illustrated in Figure 7.5d. The advantage of this method is that the sample can lie in a more conventional position on a microscope slide and it is also possible for samples to be passed under the system in micro-fluidic channels for high throughput imaging. The clear disadvantage is that of having to mount the majority of the optics vertically.

The other major variation, though again not common, is that of using dual beam illumination (Huisken and Stainier 2007) (see Figure 7.5b). Here the excitation beam is brought in from both sides of the sample to help combat loss of light through the sample and to produce a more uniform illumination across the entire field of view. While this clearly works it adds complexity as the two light sheets need to be perfectly overlapping and in the focal plane

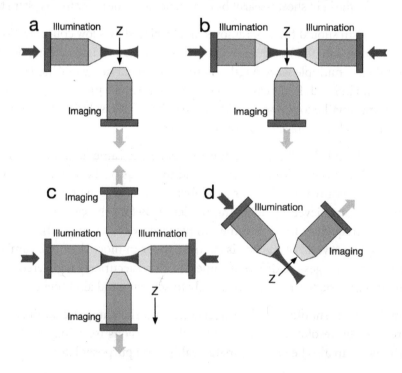

FIGURE 7.5 Alternative SPIM configurations: a) conventional SPIM, b) mSPIM with dual illumination, c) Mu-Vi SPIM with dual illumination and imaging, d) i-SPIM inverted SPIM configuration.

of the imaging objective. A further advance here is to then consider the light that is being emitted in the opposite direction to the single collection objective and thus to add a further imaging objective in this direction (de Medeiros et al. 2015) (see Figure 7.5c). Further details on these, and other, more advanced SPIM techniques are discussed in Section 7.7. All utilize the fact that the excitation light and imaging optics have different optical paths and directions, though the methods become increasingly complex to build and maintain, and require more complex data processing to obtain reliable high quality image sets.

7.5 PRACTICAL OPERATION

Having now aligned the system the actual sample mounting and operation of an SPIM system needs to be considered. As with all optical methods the overriding principle has to be to record the best quality data set in the simplest manner to provide information to answer the specific question being explored. Practicality should always be the main driver, and when SPIM is used for extended imaging periods, frequently over several days in developmental biology, the method has to be reliable. The other consideration that is important in SPIM, especially when used for extended periods, is the size of the data set that will be generated and how this will be stored, as discussed in Section 7.6.

7.5.1 Sample Mounting

In general SPIM is a technique that is used on live samples and thus keeping the sample in as natural state as possible is the main consideration. Many SPIM systems also use water-dipping objectives as they provide a long working distance (so the two orthogonal lenses do not touch) and a high numerical aperture (typically around 0.8 for the imaging objective) and thus the sample needs to be held in water. As described above, in a conventional SPIM the optics are mounted parallel with the table and thus the sample is introduced vertically into the imaging volume. One also needs high quality optical access to the sample from at least two orthogonal directions, and ideally all sides, so the sample can be orientated to obtain the optimum view. The exact imaging protocol for a specific experiment (for example the level of anaesthesia) is best determined based upon experience and a review of the most up to date literature. In the case of zebrafish, melanocytes develop at around 24–48 hours and the melanin present absorbs visible light which can lead to dark areas in the image. The formation of these can be prevented through the use of around a 0.003% solution of 1-phenyl 2-thiourea (PTU). The melanocytes will be suppressed as long as the PTU is present. This does clearly perturb the zebrafish development but not sufficiently to alter most biological function (Karlsson, von Hofsten and Olsson 2001). A further alternative is through the use of the genetically modified fish "Casper" in which the melanocyte formation has been "switched-off" (White et al. 2008). The list of mounting methods below provides an excellent starting point for many experiments but is no substitute for discussions with experienced users and experimentation for each specific application.

1. Direct Glass Pipettes and Water

    Here the sample is gently drawn into a small glass capillary with an outside diameter of around 1 to 2 mm and with as thin a wall thickness as possible. This can be achieved either through capillary action or through the application of gentle suction. Care is required to ensure no air bubbles are brought into the tube as this will drastically distort the image. Trapped bubbles can sometimes be moved to the top of the tube by gentle tapping. Once the sample is in the tube (and typically this is used for zebrafish or excised organs) the lower end can be plugged to stop the sample falling out. One needs to consider which way up the sample should be mounted as the plug can obscure the organs near the bottom of the tube.

The advantages of this method are that it requires minimal preparation and that the glass tube is highly transparent. The disadvantage is that the sample is free to move (and one can have spontaneous movement even in anaesthetized samples). In addition the tube acts like a lens as it is curved and has a higher refractive index than the water, leading to aberrations (Bourgenot et al. 2012). It is possible to obtain square "tubes", but the corners can cause problems as the sample is rotated and thus aligning the sample correctly can be difficult. There is also the inherent risk with glass that the tube may break in the imaging "tank"!

2. FEP (Fluorinated Ethylene Propylene) Tubing

This is used in a very similar way to the glass tubing but has the advantage that it has a refractive index that is closely matched to that of water. Thus although it is curved it produces little distortion in the image. Being plastic it is also easy to cut to size and does not break. The disadvantage of FEP is that it is frequently curved as it is stored on a reel. Although it has a similar refractive index as water it can scatter the light due to marks on the surfaces caused during the production process.

3. Agar Mounting

The sample can also be mounted in low melting point agar and then viewed either directly or in combination with the tube methods described above. Typically a 0.5–1% solution is used (a 0.5% solution is made from 10 mg of agarose in 2 ml of water). The agarose should be mixed with the water at around 45°C and then allowed to cool to around 37°C before the sample is introduced. The sample can be placed in the agarose and then slow sucked into a warmed tube (FEP or glass) and allowed to cool slowly over about two minutes. Rapid cooling can lead to the agarose becoming scattering.

The sample can then be viewed mounted in the tube or in some cases the sample and agarose can be carefully pushed out of the tube (using a syringe at one end) and the extruded agarose and sample mounted into the SPIM system. This has the advantage of having no glass or FEP tubing but producing a straight extrusion requires practice! The agarose method helps to stabilize the sample and minimizes the involuntary movements mentioned above.

4. Inverted SPIM; Open Mounting

In an inverted configuration the sample can be mounted in a more conventional manner on a glass slide or in grooves cast within agarose and viewed from above.

Again to limit movement the sample can be embedded fully in agarose or in some cases directed under the viewing optics using a microfluidic system.

5. Root Imaging

One specific application that requires a slightly different protocol is that of imaging living plants. Here one wishes to perturb the system as little as possible and thus if one is imaging the roots one does not want the excitation light to fall on the leaves, but one does want good growth conditions. The roots can be made to germinate in an enriched agarose gel and the leaves grown through a black cover (Maizel et al. 2011) where they are exposed to artificial sunlight. The development of the root can then be studied as the plant develops under near normal conditions, with the root in the SPIM system and the leaves above.

7.5.2 Basic Operation

Having mounted the sample and confirmed the alignment of the system the next task is to start imaging the sample. The process order below is not based around any specific SPIM system but provides some generic guidelines that can be followed. It assumes the system has been fully switched on and the cameras adjusted to settings based upon previous experience, as these are very system dependent.

1. Place the mounted sample in the imaging system.

2. In a system with transmission illumination switch this light on and adjust the sample position until something is seen. It is worth noting here that if one is only viewing a single optical section of the sample it can be difficult to determine exactly where you are viewing and thus this initial "wide field" imaging modality is very useful to help in orientating the sample.

3. Switch off the transmission illumination and activate the required sheet illumination, starting at a low power. Find the region of interest in the sample.

4. Manually determine the front and back positions of the image stack and record these as required to set the imaging volume.

5. Adjust the laser intensity and camera gain to minimize photobleaching (i.e. light exposure) but using the maximum dynamic range on the camera while keeping the noise to a minimum (do not set any camera gain to high). This adjustment is important but dependent on the sample and camera sensitivity. The guiding principle should always been to keep the illumination light as low as possible.

6. Undertake a quick scan through the sample using a look-up table on the camera (or real time histogram) that clearly illustrates regions of overexposure. Repeat step 5 to remove any saturation areas as required.

7. Set the full imaging parameters and in determining the spacing of the optical slices consider the Nyquist limit for the system. The optical sections in an SPIM system are

normally thicker than in a confocal system and thus the axial spacing of the slices can generally be larger, whilst one still obtains Nyquist resolution.

8. Check the size of the data file that will be generated and that there is sufficient storage.

9. Record the data set.

7.5.3 Correcting Common Faults

The most common faults seen in SPIM images are illustrated in Figure 7.6 and the points below are intended as a guide to identifying and correcting the problem. If you are using a commercial system then a good instruction manual should also provide a practical fault finding routine.

a. Non-common focal plane: Here the light sheet and the imaging optics are not at a common focus. This leads to an image with low light levels. One route to solve this is to switch off the light sheet and to use the transmission light to ensure it is known where the imaging cameras are focused. The position of the sample can then be adjusted to bring the front of the sample into focus. The light sheet should then be switched on, and the transmission light off, and the position of the light sheet, along the viewing axis, adjusted to maximize the intensity of the image. This routine works well with the bead sample mentioned in Section 7.4.2.

b. Skewed light sheet: The image looks bright on one side and dull on the other. This means the light sheet is passing through the sample at an angle and thus is not orthogonal to the imaging objective. By adjusting the angle of the light sheet the image can be made more uniform. If the effect is in the vertical direction then the light sheet may be tilted in the other axis.

c. Shadowing: Without a resonant mirror this is hard to overcome except by ensuring the sample is a clear as possible. It may be possible to alter the angle of the sample relative to the light sheet so that the area of interest is still in the field of view but being excited, and viewed, at a slightly different angle. Heavily pigmented regions cannot be overcome except by considering using a different excitation wavelength to avoid the absorption. If this route has to be followed this may well entail using a different fluorophore for visualization. Multiphoton excitation may also get around major absorption challenges.

d. Camera misalignment: In systems with multiple cameras then the different images when overlaid may not align perfectly. This misalignment can appear as lateral and vertical shifts in the image and also rotational shifts. These can be corrected by accurate camera alignment and in the final resort image registration during processing.

e. Saturated pixels: These appear as very bright features and as a peak at the right hand end of a conventional histogram. The detail lost here cannot be recovered and the excitation intensity should be reduced. Alternatively if the cameras are at high gain or long exposure reduce these setting.

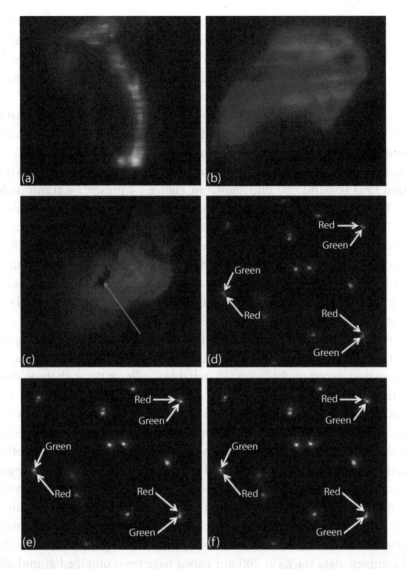

FIGURE 7.6 A collection of images illustrating common faults in SPIM: a) shadowing in the kidney – glomerulus at the top, with nephron descending, b) shadowing in liver, c) fluorescence being absorbed in the liver, d) relative rotation of red and green cameras, e) poorly focused camera, f) light sheet and imaging arm not orthogonal.

7.5.4 High Speed Imaging and Synchronization

As noted above one of the main advantages of SPIM is the speed with which the technique captures an entire optical section. This makes it the preferred choice for imaging high-speed biological events such as the beating heart, vesicle movement and calcium signalling. In the case of heart movement the process is inherently cyclic and thus one can use the repeated patterns to determine the exact movement of cells within the heart.

In one variation the SPIM system (or indeed any other high-speed imaging method) is left to free run, recording images as rapidly as possible. The images capture the movement of the heart at various points during the cycle; then using post processing the different images are "reshuffled" in time to put them into the correct movement sequence (Liebling et al. 2005). This requires a very high-speed camera and complex processing software but has been used to monitor blood flow and cell movement within the heart. An alternative is to operate the imaging in a "time gated" manner in which the imaging laser and camera are activated just before the heart is going to be at a certain point in its cycle. A single high-speed image is captured at this time point in the cycle. A second image is then captured at a slightly different time point in the cycle and this process is repeated until the full cycle has been captured in a sequence of images (Girkin and Carvalho 2018). In this method the near infrared LED (around 850 nm) shown in Figure 7.4 presents a transmission image onto a near infrared camera running at around 100 frames per second. Using real time processing after about two to three cycles the system can predict when the heart is going to be in a particular position in the beating cycle and thus fire the imaging system at the precise time to capture that cycle point. This is repeated multiple times to build up a moving image of the heart. The latter method is more accurate than the high-speed video approach but requires the ability to undertake real time image processing and control of the light sheet and camera shutter.

These methods effectively "freeze" the motion of cyclical processes. Imaging calcium, or vesicle dynamics, is not as predictable and thus requires direct high-speed imaging. In the case of calcium the interest might be within individual cells, or through three-dimensional ensembles, in particular nerves. Here even the rapid imaging of conventional SPIM is not sufficient and multiple ways have been explored to develop high-speed three-dimensional data stacks. The most successful so far is "swept confocally aligned planar excitation" or SCAPE (Hillman et al. 2015). The basic idea is shown in Figure 7.7. This method uses a single objective and the light sheet is scanned through the lens at an angle using a polygon mirror. The objective collects the fluorescence light, which is then reflected onto the camera by an adjacent element on the polygon array. This has the effect of sweeping the light sheet rapidly through the volume of the sample with only the rotation of the mirror. In the same publication this method was also demonstrated in *Drosophila*, to observe the beating heart, and complete data stacks at 200 μm cubed have been obtained at multiple frames per second.

7.5.5 High Throughput SPIM

One other area in which the ability of SPIM to image optical sections at high speed is in the development of high throughput screens for both drugs and toxicology. This is an area of increasing importance as one moves beyond a screen based upon a few cultured cells. Such screens require the ability to rapidly image one sample after another. The standard method of doing this is by moving the sample rather than the heavy optics. In an ideal world the sample might just flow past the imaging position, and this is now being achieved using microfluidic channels and SPIM. The optical configuration of a conventional SPIM could be used with the samples flowing through a tube mounted in the normal position described above. However, there are limitations here and one variation is to build an SPIM system that images from below as the samples flow along horizontally and this has been

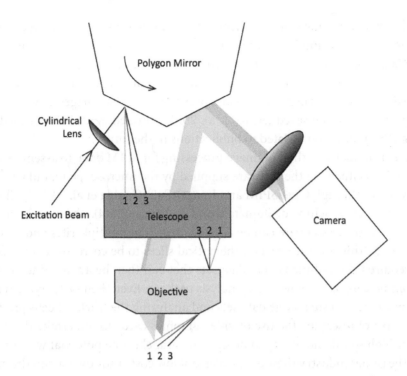

FIGURE 7.7 Simplified optical path of the swept confocally aligned planar excitation (SCAPE) microscope. (based upon Bouchard et al., *Nature Photonics*, 2015.)

achieved using a prism (McGorty et al. 2015) in a system reminiscent of the TIRF microscopes described in Chapter 12. In the case of zebrafish the samples can be aligned using features within the channel to ensure that each sample is viewed from approximately the same angle.

7.6 SPIM IMAGING PROCESSING

As discussed at the beginning of this chapter, technology, both optical and digital, is vitally important in light sheet microscopy. It has also been mentioned throughout this chapter that SPIM produces vast quantities of data that have to be stored and processed. A full detailed description of the imaging processing methods is well beyond the scope of a single chapter but the following provides a starting guide, and perhaps most importantly some of the considerations that need to be made when imaging using SPIM.

Again, before undertaking any imaging, one must consider why the images are being taken, and what will be done with the resulting data. In SPIM this may be even more crucial as the data sets can become very large very quickly. If we take a 12-bit s-VGA camera (1024 × 768 pixels) a single image is 1 Mb, and this is not a high specification camera. Thus if we then image through a sample that is 100 μm thick with a system that has an axial resolution of 1 μm at the Nyquist limit we will have 200 images or 200 Mb of data for each image stack. Although large this is not unusual for a data set from a confocal or even time-lapse wide field microscope. However, in the case of SPIM we might be tracking cell movement, which may require imaging every five minutes for 24 hours creating over 57 Gb of data. If the camera is a 5-megapixel camera at 16-bit resolution and we are imaging 200 μm (around the size of a 48 hour post fertilization zebrafish heart) this number becomes 1.27 Tb, and this is from a single experiment. Thus simply transferring the data and saving it will take considerable periods of time. It may well be that in the future raw data sets cannot be saved for a full set of experiments and only the processed images, and the associated processing details, will be saved (Amat et al. 2015; Schmied et al. 2016). This does raise interesting ethical questions related to publications in the future.

The current standard method of image processing for SPIM data, to assemble the original data set, is to either use the software supplied by the microscope manufacturer or the versions available through OpenSPIM and ImageJ/Fuji (Gualda et al. 2015; Preibisch et al. 2010; Royer et al. 2015; Schmied, Stamataki and Tomancak 2014). The most up to date version of the open source software can be downloaded from multiple sites and is constantly being improved. This software enables the optical slices to be correctly assembled to produce the required three-dimensional data sets that can then be analyzed using the most suitable commercial, or home-written, analysis package. Even then systems are not used to following movement in such large data sets and analyzing the tracks of cells produced is a very active area of research. The use of video graphics cards to undertake the processing due to their high speed and ability to process in parallel is one potential way forward and thanks to the games industry these are available at low cost. Thus once again the message is to think carefully what data will be required at the end of the experimental work and how this will be handled – before starting the imaging!

7.7 ADVANCED SPIM METHODS

There are multiple reviews available for light sheet microscopy and it is suggested that the reader searches the recent literature to see the most advanced techniques. However, a short summary is given below and these current methods are also summarized in a recent table (Girkin and Carvalho 2018).

1. Digital Light Sheet Microscopy (DSLM)

   Here the sheet of light is created using a thin beam of light that is scanned rapidly in the vertical direction to produce a sheet of light when time averaged. The main advantage is that the thin line of light scanned vertically through the sample is synchronized to a "rolling shutter" on the camera. The shutter only activates a single line of pixels at a time and thus light that might be scattered by the sample, that would normally provide background to the other pixels, is rejected. In effect one has a line-scanned confocal system to reject out of plane, and indeed in plane light, outside the part of the sample being imaged (Keller et al. 2008).

2. HiLo Light Sheet Microscopy

   This method also aims to improve the optical sectioning and contrast, by increasing the rejection of the out of focus light. The sample is initially imaged using a conventional plane sheet of light and then with a structured light sheet (a series of excited and non-excited lines). The two images are then combined using an algorithm utilizing the high (Hi) and low (Lo) spatial frequency information from the two images (Mertz and Kim 2010)

3. Multiview SPIM (MuVi-SPIM)

This method uses both the two-beam illumination described earlier and also two objectives to collect light from both sides of the sample (Figure 7.5c). The two beams can also be at different focal planes each imaged into one camera. Thus by interleaving the illumination and camera shutters complete data sets can be recorded at twice the speed. In addition deeper imaging is enabled as one can see both sides of the sample (de Medeiros et al. 2015).

4. Non-Linear or Multiphoton SPIM

Here the light sheet is produced using a near infrared, ultra-short pulsed laser such that two photons are absorbed to excite the fluorescence further localizing the excitation (see Chapter 8). The light sheet is again generated as a scanned line and the technique allows greater penetration of light into the sample. However, the fluorescence still has to be collected through the sample thus the gain is not as great as that for conventional beam scanned multiphoton microscopy (Fahrbach et al. 2013)

5. Exotic Beam Microscopy (Bessel and Airy Beam)

In order to produce a light sheet that has a more uniform thickness over a larger area a range of different light beams have been used. These beams have a lower divergence but have higher intensity side lobes, which means they excite fluorescence outside the main focal plane. Once the images have been recorded they need to be post processed using de-convolution software to remove these out of plane artefacts (Planchon et al. 2011; Vettenburg et al. 2014).

6. Super-Resolution Light Sheet Microscopy (Lattice Sheet Microscopy)

The full details of super-resolution imaging are presented in Chapter 12 but the various methods enable the diffraction limit of an optical system to be defeated, leading to optical images with resolutions of significantly less than 100 nm. In the case of light sheet methods an array of light beams is produced such that it creates a three-dimensional lattice of light spots within the sample. These are then imaged, and the centre of each found; then another lattice is produced until eventually the entire volume has been illuminated. The centres of each point of fluorescence are then combined to provide an image. The process is slow but does produce images with significantly higher resolution (Chen et al. 2014).

7.8 WHAT IS SPIM GOOD FOR AND LIMITATIONS

As a microscopy technique, light sheet methods as currently operated, are new and the technical developments in the field are currently very rapid. The main advantages of the method are, however, clear. If one requires three-dimensional images of an intact, living, basically transparent sample, SPIM is probably the first method to consider. If one then adds to this the desire to image at higher speed in three dimensions then it is without doubt the route to explore. In all light microscopy methods the sample has to be illuminated and while for fixed, dead samples the light is not normally harmful (though you may have a little photo-bleaching), for anything that is living this has to be a consideration. Light sheet microscopy delivers the lowest light dose to the sample of any three-dimensional imaging method making it an excellent choice for extending imaging of samples. In taking a single optical section in one image it is also the fastest method and thus ideally suited for dynamic imaging. The resolution of the image can be diffraction limited in the main plane, however. In the optical axis the thickness of the slice is larger than one would achieve using a well-adjusted confocal microscope with a high numerical aperture objective. There are complex ways around this limitation but they are not simple to implement.

The main disadvantage, beyond the resolution discussed above, is that the sample needs to be transparent and not highly scattering. The technique has been used with samples that have been chemically cleared, with outstanding results; however, those samples are dead and have been perturbed in many ways. The use of PTU (discussed in Section 7.5) to clear the sample in the case of zebrafish is also known to alter the development but the advent of "Casper" fish where the melanocytes have been genetically removed offers another alternative (White et al. 2008). The size of the data sets that can be produced can also be a limiting factor but this should not really be the reason why the method is not selected if everything else points towards SPIM.

REFERENCES

Amat, F., B. Höckendorf, Y. Wan, W. C. Lemon, K. McDole and P. J. Keller. 2015. "Efficient Processing and Analysis of Large-Scale Light-Sheet Microscopy Data". *Nature Protocols* 10(11): 1679–96. www.nature.com/doifinder/10.1038/nprot.2015.111%5Cnhttp://www.nature.com.gate2.inist.fr/nprot/journal/v10/n11/full/nprot.2015.111.html.

Bourgenot, C., C. D. Saunter, J. M. Taylor, J. M. Girkin and G. D. Love. 2012. "3D Adaptive Optics in a Light Sheet Microscope". *Optics Express* 20(12): 13252–61. www.ncbi.nlm.nih.gov/pubmed/22714353.

Chen, B.-C., W. R. Legant, K. Wang, L. Shao, D. E. Milkie, M. W. Davidson, C. Janetopoulos, et al. 2014. "Lattice Light-Sheet Microscopy: Imaging Molecules to Embryos at High Spatiotemporal Resolution". *Science* 346(6208): 439.

Engelbrecht, C. J. and E. H. K. Stelzer. 2006. "Resolution Enhancement in a Light-Sheet-Based Microscope (SPIM)". *Optics Letters* 31(10): 1477–9.

Fahrbach, F. O., V. Gurchenkov, K. Alessandri, P. Nassoy and A. Rohrback. 2013. "Light-Sheet Microscopy in Thick Media Using Scanned Bessel Beams and Two-Photon Fluorescence Excitation". *Optics Express* 21(11): 13824. www.opticsinfobase.org/abstract.cfm?URI=oe-21-11-13824 (June 1).

Girkin, J. M. and M. T. Carvalho. 2018 "The Light Sheet Microscopy Revolution". *Journal of Optics* 20: 053002.

Greger, K., J. Swoger and E. H. K. Stelzer. 2007. "Basic Building Units and Properties of a Fluorescence Single Plane Illumination Microscope". *Review of Scientific Instruments* 78(2): 23705. http://aip.scitation.org/doi/10.1063/1.2428277.

Gualda, E. J., H. Pereira, T. Vale, M. F. Estrada, C. Brito and N. Moreno. 2015. "SPIM-Fluid: Open Source Light-Sheet Based Platform for High-Throughput Imaging". *Biomedical Optics Express* 6(11): 4447. www.osapublishing.org/viewmedia.cfm?uri=boe-6-11-4447&seq=0&html=true.

Hillman, E.M.C., M. B. Bouchard, V. Voleti, C. S. Mendes, C. Lacefield, V. George, W. B. Grueber, et al. 2015. "Swept Confocally-Aligned Planar Excitation (SCAPE) Microscopy for High-Speed Volumetric Imaging of Behaving Organisms". *Nature Photonics* 9(2): 113–19. http://dx.doi.org/10.1038/nphoton.2014.323.

Huisken, J., J. Swoger, F. Del Bene, J. Wittbrot, E. H. Stelzer. 2004. "Optical Sectioning Deep inside Live Embryos by Selective Plane Illumination Microscopy". *Science (New York)* 305(5686): 1007–9. www.ncbi.nlm.nih.gov/pubmed/15310904.

Huisken, J. and D. Y. Stainier. 2007. "Even Fluorescence Excitation by Multidirectional Selective Plane Illumination Microscopy (mSPIM)". *Optics Letters* 32(17): 2608–10. www.ncbi.nlm.nih.gov/pubmed/17767321.

Karlsson, J., J. von Hofsten, and P. E. Olsson. 2001. "Generating Transparent Zebrafish: A Refined Method to Improve Detection of Gene Expression during Embryonic Development". *Marine Biotechnology* 3(6): 522–7. www.ncbi.nlm.nih.gov/pubmed/14961324.

Keller, P. J., A. D. Schmidt, J. Wittbrodt and E. H. K. Stelzer. 2008. "Reconstruction of Zebrafish Early Embryonic Development by Scanned Light Sheet Microscopy". *Science (New York)* 322(5904): 1065–9. http://science.sciencemag.org/content/322/5904/1065.abstract.

Liebling, M., A. S. Forouhar, M. Gharib, S. E. Fraser, M. E. Dickinson. 2005. "Four-Dimensional Cardiac Imaging in Living Embryos via Postacquisition Synchronization of Nongated Slice Sequences". *Journal of Biomedical Optics* 10(5): 54001. www.ncbi.nlm.nih.gov/pubmed/16292961.

Maizel, A., D. von Wangenheim, F. Federici, J. Haseloff and E. H. Stelzer. 2011. "High-Resolution Live Imaging of Plant Growth in near Physiological Bright Conditions Using Light Sheet Fluorescence Microscopy". *The Plant Journal* 68(2): 377–85. http://doi.wiley.com/10.1111/j.1365-313X.2011.04692.x.

McGorty, R., H. Liu, D. Kamiyama, Z. Dong, S. Guo and B. Huang. 2015. "Open-Top Selective Plane Illumination Microscope for Conventionally Mounted Specimens". *Optics Express* 23(12): 16142. www.osapublishing.org/abstract.cfm?URI=oe-23-12-16142.

de Medeiros, G., N. Norlin, S. Gunther, M. Albert, L. Panavaite, U. M. Fiuza, F. Peri, et al. 2015. "Confocal Multiview Light-Sheet Microscopy". *Nature Communications* 6: 8881. www.nature.com/ncomms/2015/151125/ncomms9881/abs/ncomms9881.html.

Mertz, J. and J. Kim. 2010. "Scanning Light-Sheet Microscopy in the Whole Mouse Brain with HiLo Background Rejection". *Journal of Biomedical Optics* 15(1): 16027. www.ncbi.nlm.nih.gov/pubmed/20210471.

Pitrone, P. G., J. Schindelin, L. Stuyvenberg, S. Preibisch, M. Weber, K. W. Eliceiri, J. Huisken, et al. 2013. "OpenSPIM: An Open-Access Light-Sheet Microscopy Platform". *Nature Methods* 10: 598–9. www.nature.com/doifinder/10.1038/nmeth.2507 (June 10).

Planchon, T. A., L. Gao, D. E. Milkie, M. W. Davidson, J. A. Galbraith, C. G. Galbraith and E. Betzig. 2011. "Rapid Three-Dimensional Isotropic Imaging of Living Cells Using Bessel Beam Plane Illumination". *Nature Methods* 8(5): 417–23.

Preibisch, S., S. Saalfeld, J. Schindelin and P. Tomancak. 2010. "Software for Bead-Based Registration of Selective Plane Illumination Microscopy Data". *Nature Methods* 7(6): 418–19.

Royer, L. A., M. Weigert, U. Günther, N. Magheli, F. Jug, I. F. Sbalzarini and E. W. Myers. 2015. "ClearVolume: Open-Source Live 3D Visualization for Light-Sheet Microscopy". *Nature Methods* 12(6): 480–1. www.nature.com/doifinder/10.1038/nmeth.3372.

Schmied, C., E. Stamataki and P. Tomancak. 2014. "Open-Source Solutions for SPIMage Processing". In *Methods in Cell Biology*, 505–29. http://linkinghub.elsevier.com/retrieve/pii/B9780124201385000276.

Schmied, C., P. Steinbach, T. Pietzsch, S. Preibisch and P. Tomancak. 2016. "An Automated Workflow for Parallel Processing of Large Multiview SPIM Recordings". *Bioinformatics* 32(7): 1112–14. https://academic.oup.com/bioinformatics/article-lookup/doi/10.1093/bioinformatics/btv706.

Siedentopf, H. and R. Zsigmondy. 1902. "Uber Sichtbarmachung und Größenbestimmung Ultramikoskopischer Teilchen, mit Besonderer Anwendung auf Goldrubingläser". *Annalen der Physik* 315(1): 1–39. http://doi.wiley.com/10.1002/andp.19023150102.

Stelzer, E. H. K. and S. Lindek. 1994. "Fundamental Reduction of the Observation Volume in Far-Field Light Microscopy by Detection Orthogonal to the Illumination Axis: Confocal Theta Microscopy". *Optics Communications* 111: 536–47.

Taylor, J. M., J. M. Girkin and G. D. Love. 2012. "High-Resolution 3D Optical Microscopy inside the Beating Zebrafish Heart Using Prospective Optical Gating". *Biomedical Optics Express* 3(12): 314–21.

Vettenburg, T., H. I. Dalgamo, J. Nylk, C. Coll-Lladó, D. E. Ferrier, T. Čižmár, F. J. Gunn-Moore and K. Dholakia. 2014. "Light-Sheet Microscopy Using an Airy Beam". *Nature Methods* 11(5): 541–4.

Voie, A. H., D. H. Burns and F. A. Spelman. 1993. "Orthogonal-Plane Fluorescence Optical Sectioning: Three-Dimensional Imaging of Macroscopic Biological Specimens". *Journal of Microscopy* 170(3): 229–36. http://doi.wiley.com/10.1111/j.1365-2818.1993.tb03346.x.

White, R. M., A. Sessa, C. Burke, T. Bowman, J. LeBlanc, C. Ceol, C. Bourque, et al. 2008. "Transparent Adult Zebrafish as a Tool for In Vivo Transplantation Analysis". *Cell Stem Cell* 2(2): 183–9.

Wu, Y., A. Ghitani, R. Christensen, A. Santella, Z. Du, G. Rondeau, Z. Bao, et al. 2011. "Inverted Selective Plane Illumination Microscopy (iSPIM) Enables Coupled Cell Identity Lineaging and Neurodevelopmental Imaging in Caenorhabditis Elegans". *Proceedings of the National Academy of Sciences* 108(43): 17708–13. www.pnas.org/cgi/doi/10.1073/pnas.1108494108.

CHAPTER 8

Multiphoton Fluorescence Microscopy

IN MANY WAYS MULTIPHOTON microscopy epitomizes one of the core messages of the book, that of imaging a sample in three dimensions with minimal perturbation to the system under observation. Since the first demonstration of multiphoton microscopy (Denk, Strickler and Webb 1990) the technique has rapidly been accepted as the method of choice for fluorescence-based imaging at depth with high spatial resolution. The original publication used two photon excitation and this demonstration of non-linear microscopy led to an explosion of methods which are described in later chapters, including harmonic microscopy (Chapter 9) and non-linear Raman microscopy (Chapter 10). As will emerge throughout this chapter the growth in the requirement for complex, ultra-short pulse laser systems, needed as the excitation source, helped to stimulate a renaissance in collaborations between physical and life scientists.

This chapter initially introduces the main physical concepts behind multiphoton microscopy and stresses when it is likely to be the method of choice for an imaging application. This is followed by the practical considerations in operating a multiphoton microscope, linking back to Chapter 5 for the beam scanning methods. As most people who approach multiphoton microscopy will have had experience with a confocal system, the chapter then looks at the best approach to using a multiphoton system based upon previous confocal experience. This includes some of the compromises that might need to be made in the instrument settings to obtain the most useful data sets. The chapter then ends with some details on the miniature endoscopes being developed for *in vivo* microscopy, and even moving towards clinical applications. This section also includes simple ways of quantifying the two photon light dose in a sample and how to minimize any potential unwanted side effects. In order to keep the nomenclature simple, the term *"multiphoton microscopy"* will be used throughout the chapter to refer to fluorescence excitation by two, or more photons.

8.1 INTRODUCTION TO MULTIPHOTON EXCITATION

In a fluorescently labelled sample, in order to obtain three-dimensional images with high resolution one really has two options. One route is to localize the detection from a single small volume within the sample, which is that used in a confocal microscope. A second option is to localize the excitation, which is the one adopted by all the non-linear imaging methods. The most important common feature in all these methods is that the material being excited fluoresces in a non-linear manner, with the excitation intensity leading to localized excitation due to the focusing of the excitation beam. It is only at the focus of the beam that the excitation intensity is high enough to generate detectable fluorescence.

The core concept of multiphoton fluorescence imaging is straightforward and the basic physical theory was first published in an article by Goeppert-Mayer in 1931, for which she was awarded the Nobel Prize for Physics in 1963. Using the then recently developed theory of quantum mechanics, she postulated, and demonstrated mathematically, that it was possible for electrons to move between different quantum mechanical energy levels through the absorption, or in fact emission, of two or more quanta of light (photons). The total energy of the photons absorbed was found to be the equivalent of a single photon at a shorter wavelength. As the probability of such a multiple photon event happening is very low (compared to single photon transitions) this effect was very much viewed as a mathematical, or quantum mechanical, curiosity. The advent of the laser changed this position and multiphoton events were used as a core technique in high-resolution atomic spectroscopy totally transforming the field. Non-linear excitation for microscopy was first mentioned by Wilson and Sheppard in 1984 though it was a further six years before the method was demonstrated (Denk, Strickler and Webb 1990). Although the concept of non-linear microscopy had been discussed before, and indeed demonstrated in non-biological systems, the paper by Denk and colleagues showed that non-linear imaging, with minimal damage to the sample, was possible.

In multiphoton microscopy the core method is to create a small, very intense focus of light, at twice the normal excitation wavelength, within the sample. This highly localized intensity leads to a significant probability that two, or more, photons will be absorbed at the same time leading to fluorescence. This process is illustrated in Figure 8.1. In Figure 8.1a light at a short wavelength (say around 405 nm) creates fluorescence through conventional absorption in the fluorophores present. This produces the classic fluorescence profile as the light is focused into the sample. However, in Figure 8.1b two infrared photons (say at 810 nm) are absorbed simultaneously where the light intensity is high enough. This will only be the case right at the focus of the light beam and thus one has localized excitation within the sample. The resulting fluorescent signal can then only have come from the focal point of the excitation beam and by scanning this point across the sample an optical slice can be constructed. The localized excitation inherently leads to optical sectioning, and by then scanning at different depths within the sample a three-dimensional data set can be recorded.

The main advantages to the researcher of the multiphoton fluorescence technique is that it enables imaging at significantly greater depths within samples, along with increased

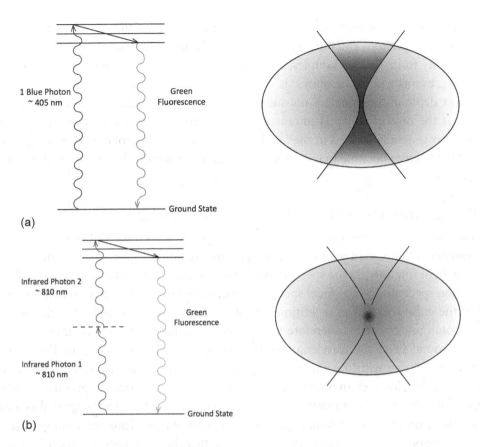

1 Blue Photon
~ 405 nm

Green
Fluorescence

Ground State

(a)

Infrared Photon 2
~ 810 nm

Green
Fluorescence

Infrared Photon 1
~ 810 nm

Ground State

(b)

FIGURE 8.1   One (a) and two (b) photon excitation and the associated fluorescence intensity.

viability of live specimens, providing the conditions are correct. This ability to image more deeply arises from two physical principles. The excitation wavelength is typically in the near infrared (~720–950 nm from a titanium sapphire (Ti:Sapphire) laser); this wavelength is not linearly absorbed by most biological tissue and scattering is reduced at these longer wavelengths compared to visible excitation. The lack of linear absorption means that as one moves more deeply into the sample one is able to increase the laser power and maintain a high peak power. This is despite the loss of a few photons due to scattering or aberration caused by the optical properties of the tissue being imaged. In recent years the use of three photons, at around 1500 nm, has started to become the imaging method of choice for deep brain imaging due to the significantly lower scattering at even longer wavelengths.

The second, perhaps more important principle, is that as a result of the localization of the excitation, one can collect all of the fluorescent light. In the other main optical section-ing microscope, the confocal microscope, the fluorescent light has to pass back through the instrument and subsequently a pinhole to obtain the required sectioning ability. Thus any scattering of the fluorescent light as it emerges from the sample will mean signal is lost as the light, although perhaps emanating from the correct focal depth, will be rejected by the pinhole. In the case of the multiphoton system all the fluorescent light can be utilized to build up the image without the need to be descanned and pass through optical apertures as all the signal photons can only have come from the focal volume.

The ability to image live samples, with minimal damage, is possible as the number of phototoxic events within the sample is significantly reduced under the correct conditions. The entire depth of the sample is not subjected to potentially damaging blue or ultraviolet light as is the case in confocal microscopy. Any damage that is caused by multiphoton processes is generally limited to the focal plane. By careful control of the imaging conditions it is therefore possible to obtain images from deep within the sample with minimal perturbation.

8.1.1 Requirements on Light Sources for Multiphoton Excitation

We now need to consider the technology challenges posed by multiphoton microscopy. The number of fluorescent photons, for two photon excitation, depends on: 1) the square of the peak light intensity, 2) the spectral overlap of the excitation light with the two photon absorption spectrum, 3) the focal spot size (determined by the lens NA and wavelength), and 4) the pulse length and repetition rate of the laser. The main technical challenge in multiphoton microscopy is therefore to produce a localized, very intense laser spot with light at around twice the wavelength of the normal one photon excitation. Producing a small focal volume within a microscope is routine through the use of high numerical aperture lenses. However, in order to achieve suitable excitation rates to produce sufficient fluorescence photons a laser power of about 1 kW is required. If this were applied as a continuous laser beam then the sample would simply boil or melt. Thus one requires a short, intense pulse of light, but not too intense to cause either thermal (generally a slow continuous wave effect) or explosive (Q-switched or nanosecond pulse) damage to the sample. The lack of such suitable sources was the main reason for the delay in the demonstration, and subsequent adoption ,of the method.

A number of factors affect the size of the fluorescent signal but most notably the number of excitation events. For a pulsed laser this is given by

$$N = I_{peak}^2 \delta \left(\frac{\pi \times NA^2}{h \times c \times \lambda} \right)^2$$

where $I_{peak} = \dfrac{P_{avg}}{v \times \tau_p}$.

Here I_{peak} is the peak intensity, P_{avg} the average power on the sample, τ_p, the pulse width and v the repetition rate of the laser; NA is the numerical aperture of the objective lens, λ the laser wavelength and h, c and π have their normal physical meanings, while δ is the two photon absorption cross-section for the transition at the excitation wavelength and is typically around 10^{-50} cm^4 s/photon/molecule. For practical implementation of this equation in microscopy the repetition rate is limited by the considerations of the fluorescence lifetime of the fluorophore and also the speed with which one wishes to image (a laser at 1 Hz may have a low repetition rate giving a high N but scanning a 512×512 pixel image would not be practical).

As discussed in Chapter 6 the time between pulses should be around five times the fluorescence lifetime. For a typical fluorophore the lifetime is around 3 ns and thus a repetition frequency of 80 MHz (12.5 ns pulse separation) is well suited to the task. This is close to the frequency of the first ultra-short pulse lasers, though more recent designs have stayed with this figure for these technical reasons.

The source should have some ability to be tuned in wavelength but as molecular absorption profiles are generally very broad, an exact wavelength match is therefore not required. In many laboratories the laser wavelength is altered very little for different fluorophores and if they were commercially available, a series of low-cost, fixed wavelength sources might well be a practical solution to the excitation source for multiphoton microscopy. Section 8.2.1 describes some practical guidance on source selection and operating parameters and for more information on suitable lasers for multiphoton microscopy the reader is referred to other works by the author (Girkin and McConnell 2005; Girkin 2008). In summary, however, one is looking for an ultra-short pulse laser with pulses of around 100–200 fs, a repetition rate of 50–80 MHz, average power of at least 200 mW (to achieve a maximum power on the sample of around 50 mW) and operation from around 750 to 950 nm, though this might be as a series of discrete wavelengths rather than having tuning over the entire range. As mentioned above there is now a growing interest in the use of longer wavelength sources, around 1500 nm, for three photon excitation in highly scattering samples such as the brain.

8.1.2 Requirements on Photon Detection for Multiphoton Microscopy

A critical feature in all microscopy is the collection of the signal photons but this is perhaps more important in non-linear microscopy. This leads to two important requirements in the detection of the precious signal photons. The first relates to the microscope objective, and the second to the beam scanning optical system. In the case of the objective we require a high NA to produce a small excitation volume, as described in Chapter 2. However, compared to confocal microscopy we have one other major consideration. Typically the excitation wavelength is now very much longer than the detection wavelength and this means that chromatic aberration can become important. Objectives are specially designed, as was discussed in Chapter 3, to be achromatic over a broad range of visible wavelengths (typically 405–647 nm) but we are now exciting at perhaps 920 nm and wanting to detect light at 560 nm (in the case of green fluorescent protein for example). This means that broadband chromatic operation is an important consideration in multiphoton microscopy. Typically with multiphoton imaging we are also aiming to image more deeply within a sample (over 200 μm) and many high NA oil objectives do not have sufficiently long working distances. Thus the best lens may not be the high NA oil objectives frequently used for confocal microscopy. Further details are described in Section 8.2.4 on the parameters to consider in selecting the most suitable imaging objective.

The chromatic aberration present both within the lens, and potentially other optics of the system, means that signal can be lost as the returned light will not be passing through the scanning system in the manner in which it was designed. However, non-linear excitation has one crucial feature which helps here. As the excitation is localized in space we can collect all the fluorescent photons as they can only have come from our excitation focal volume and a pinhole is not required to reject out of focus emission. In a conventional beam scanning confocal imaging system the light has to travel back through the optical path to be descanned before passing through the confocal aperture and reaching the detector. In a typical scanned system (as described earlier in Chapter 5 and Figure 5.2) this can require around five lenses, two mirrored surfaces and two dichromatic filters. Even with 95% transmission, or reflection, on all surfaces one could only detect at best 75% of the emitted light, assuming no other losses through transmissive optics. In practice this figure is often nearer 50% even with the best coatings.

As we do not need to pass through a pinhole in multiphoton microscopy, we do not need to de-scan the signal light beam and hence the detector can be mounted close behind the objective, as shown in Figure 8.2. Indeed fluorescence is emitted in all directions and thus

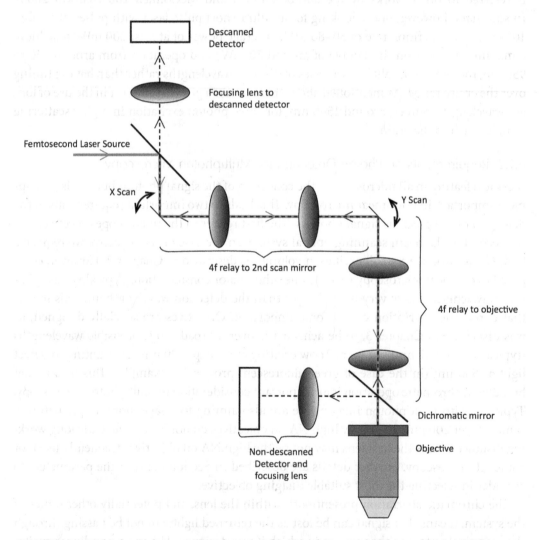

FIGURE 8.2   Multiphoton beam scanned system showing descanned and non-descanned detectors.

half of the light travels forward through the sample. In certain samples it is also possible to collect this forward emitted light and combine all the fluorescently detected light for improved signal to noise.

The actual detectors used for multiphoton microscopy are the same as those used in confocal microscopy and described in Chapter 5. There is, however, one other consideration that should be made with regard to the detectors. Some light sensitive devices, in particular silicon avalanche photodiodes and some of the faster semiconductor devices, have a very small active area where photons are detected. In a confocal system this is not a problem as the light will have been collected and focused through a pinhole and so can subsequently be directed onto the small sensitive area. In a non-descanned system the beam may move around slightly and thus if a small area detector is used, even with focusing optics, the signal may not be uniform across the entire image. It is thus preferable to user larger detectors, or sophisticated optics, to overcome any beam movement. The larger detectors also mean that fluorescence photons, scattered as they leave the sample, can be detected.

As a simple summary on the light collection and detection, use a high NA lens both for the resolution it provides, as well as the high light collection efficiency, and then detect the fluorescent photons as soon as possible in the optical system. It is also worthwhile noting here that non-descanned detectors do have one minor drawback. They are sensitive to light entering the system from any path and it is not unknown in laboratories using this detection method to have to cover all sources of light in the room including indicator LEDs, computer screens and emergency exit signs.

8.2 PRACTICAL MULTIPHOTON MICROSCOPY

Most of multiphoton microscopes are based upon beam scanned confocal systems and a typical system, illustrated in Figure 8.2, is based upon the earlier confocal instrumentation described in Chapter 5. In order to maximize the performance of a multiphoton microscope we will now look at the main variables when undertaking multiphoton microscopy.

8.2.1 Wavelength

Clearly the first requirement for the wavelength at which to operate the laser has to be that which will excite the fluorophore being used in the biological preparation. The second requirement is that it should deliver the minimal perturbation to the sample. In the case of the near infrared excitation used in multiphoton microscopy this means minimizing the direct absorption, and hence heating, in the sample. In some cases this can be water absorption as one moves above about 950 nm.

Typically an excellent wavelength to start imaging is with the excitation source at slightly shorter than twice the single photon excitation wavelength (half the photon energy). However, it may be preferable not to always tune the laser to the peak of the absorption of the fluorophore. Many fluorophores have a wide absorption band, some with a full width half maximum absorption of greater 100 nm. This can be used to the benefit of the biology being undertaken to reduce the unwanted effects of the imaging process, for example de-tuning from the peak wavelength to minimize direct absorption in water.

The actual wavelength used for the imaging probably has the largest single effect on the long-term viability of a live sample. Each wavelength and each sample has a certain toxicity threshold that can only be determined by experimental assay, but a general consensus is growing that the longer the wavelength that can be used the better for the sample. Like the fluorescence excitation the toxicity is localized in the sample, but in the focal plane the toxicity can be high. Once the toxic threshold has been exceeded in terms of peak power at a given wavelength the sample viability will deteriorate rapidly. It has been shown that the toxicity is related to the actual two photon excitation events in the UVA to blue (350–500 nm) portion of the optical spectrum. The number of excitation events are a complex combination of the sample, wavelength, pulse width, pulse shape, peak power and repetition rate.

Table 8.1 lists some of the commonly used fluorophores and a suitable starting wavelength for excitation. In selecting the wavelength it should also be borne in mind that between 850 and 950 nm many microscope objectives have a dip in their optical transmission. A consequence of this is that as the laser wavelength is tuned the fluorescent image may dim. This could be due to a fall in power on the sample rather than moving away from the peak of the absorption curve. The excitation wavelength may also be selected such that two or more fluorophores are excited simultaneously though each is away from its peak excitation. Careful controlled experiments should be undertaken to determine the correct operating parameters if the sample viability is important to the experimental result.

8.2.2 Pulse Width and Dispersion Compensation

Although multiphoton imaging has been demonstrated using continuous laser excitation (Hanninen, Soni and Hell 1994) the vast majority of imaging is undertaken with ultra-short pulse lasers. The general definition of an ultra-short pulse laser is one with a pulse length of less than 1 ps, and those used in multiphoton microscopy are typically around 100–200 fs. For a given pulse repetition rate, as illustrated in Figure 8.3, the shorter the pulse the higher the peak power, for a constant average power. As described above it is

TABLE 8.1 A Selection of Widely Used Fluorophores with the Suggested Two Photon Excitation Wavelength

Fluorophore	1 Photon Excitation /nm	2 Photon Excitation /nm	Peak Emission /nm
DAPI	350	800 (2P) 1047 (3p)	470
EGFP	470	900–920	510
Alexa 488	488	830–850	525
FITC	490	830–850	520
YFP	525	>950, 1047	540
Propidium iodide	536	>900, 1047	623
FM4-64	543	1047	640
Cy3	554	1047	568
Texas red	596	1047	620

Note: These are suggested wavelengths and may not be the most suitable for every system. Some consideration has been given to the lens transmission and laser performance in addition to the absorption profile of the fluorophore.

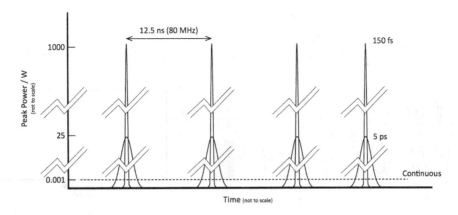

FIGURE 8.3   Ultra-short pulse laser definitions.

the peak power that dominates the excitation rate and thus the maximum signal possible. The conclusion from this would appear to be that one would like to make the pulses as short as possible. However, this is not the case in practical systems. As the pulse becomes shorter and shorter two things happen which are not advantageous for minimal perturbation imaging. First, the peak power rises, which is clearly positive. However, there is also an increased risk of the simultaneous absorption of three, or more, photons leading to the multiphoton equivalent of ultraviolet damage, or even ablation of the sample. The risk of such effects generally increases for pulses shorter than 100 fs.

Lasers are normally considered to emit a continuous output of monochromatic light, but for a short pulse laser the output contains a wide range of wavelengths (frequencies) in a repetitive series of bursts. In a conventional laser cavity only certain "modes" are allowed as the waves circulating within the cavity must fulfil a mathematical formula to ensure stable and constant feedback (Figure 8.4a). In such a laser a number of these modes oscillate, but there is no relationship between the various frequencies. This is similar to a number of church bells all ringing together, each one operated independently from the next with the result being a steady cacophony. If, however, the allowed modes within the laser cavity can be forced to oscillate with a constant relationship between themselves, they will add up at one point to produce a very intense pulse (Figure 8.4b), which can circulate around the cavity. The more modes (wavelengths) that can be made to oscillate together the shorter the pulse and the broader the optical spectrum. For a typical 100 fs pulse one has a spectral output of around 20 nm, which is large compared to the 0.01 nm output from a typical argon ion laser. In the church bell analogy when they are all forced to swing together one hears a series of intense chords with silence in between. The process of forcing the various cavity modes to operate together is known as mode-locking.

In the most commonly used ultra-short pulse laser, the Ti:Sapphire laser, the cavity modes are locked using a technique known as Kerr lens mode-locking. This is based upon the optical Kerr effect first reported by John Kerr in Glasgow in 1875. He observed that when an electric field is applied to a transparent material the optical properties of the material could be altered. The effect depends on the square of the electric field and induces a refractive index change varying on the intensity of the light. In a Kerr lens mode-locked laser the optical cavity of the laser only operates at low loss when such a lens is induced in the Ti:Sapphire

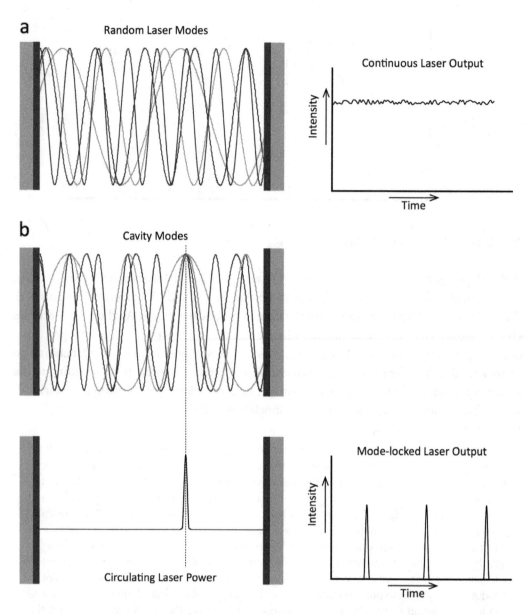

FIGURE 8.4 Laser modes and output power in a) a continuous wave laser, b) a mode-locked laser.

crystal; without this the loss in the cavity is high and this encourages the laser modes to lock together.

A typical configuration for a Ti:Sapphire laser is illustrated in Figure 8.5. The output from the pump laser is focused into the Ti:Sapphire crystal and the laser cavity is then formed using external optics around the crystal. Generally the Ti:Sapphire crystal is water cooled to reduce the thermal induced lensing and ensure stable operation. To induce mode-locking within the cavity via the Kerr lens route a disturbance is need so that the laser output develops a slightly higher peak power. One commercial company (Coherent) achieve this using a pair of glass plates that wobble when the system is not mode-locked, inducing a perturbation into the

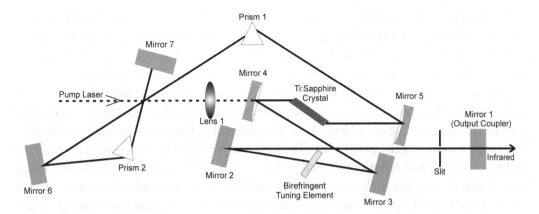

FIGURE 8.5 A typical layouts for a commercial Ti:Sapphire laser (Figure shows Mira™ courtesy of Coherent Inc.).

cavity ensuring the development of a Kerr lens in the system. A slit is then used such that only the laser mode with the induced Kerr lens can circulate through the cavity (in a similar way to a confocal pinhole only allowing the detectors to see light from the focus of an objective lens). Once mode-locked operation has been established the glass plates are held stationary.

In an alternative formulation (Spectra Physics) the glass plates are replaced by an acousto-optic modulator that gives a constant "kick" to the system ensuring that one pulse has a preference over the random modes of the laser. This "seed" is then fed into the main cavity and the induced optical effect in the laser ensures continual mode-locked operation. This technique is known as regenerative mode-locking; however, the actual short pulse generation is due to the non-linear effects induced in the Ti:Sapphire crystal resulting from the high peak powers, not the acousto-optic modulator.

According to the Heisenberg uncertainty principle energy and time are linked so that the error, or spread in energy, is linked with the uncertainty in time.

$$\Delta t \Delta E \geq \frac{\hbar}{2}$$

Thus as the pulse length, Δt, becomes shorter as the energy spread, ΔE, becomes larger. This spread in energy appears as a broadening of the wavelength of the light, which can be calculated knowing the centre wavelength and pulse length from

$$\Delta \lambda = \frac{\lambda^2}{4 \pi c \Delta t}.$$

There is also an optical effect that militates against very short pulses. As the pulse becomes shorter in time, it increases its spectral bandwidth, for instance for 100 fs pulses, around 800 nm, the pulse will have a spectral width of around 20 nm. As the pulse becomes shorter this spectral bandwidth increases rapidly. As the pulses travel through the optical elements in a microscope the shorter wavelength light is slowed relative to the red spectral portion due to the changing refractive index with wavelength (dispersion of the glass). This means that the blue light is delayed relative to the red light in the pulse, stretching the pulse length. The level of pulse stretching depends on the initial spectral width of the pulse.

In the case of the 800 nm, 100 fs pulses these have a pulse length of around 200–300 fs when they reach the focus of a standard optical microscope. This lowers the peak power by a factor of two. If 50 fs pulses were used they would have a pulse length of around 600 fs on the sample, unless special optical methods are employed to counteract this pulse lengthening.

Another consequence of the use of femtosecond pulses is that, in general, they are not compatible with delivery through an optical fibre. In the first instance the chromatic dispersion effects described above become very significant in the case of an optical fibre even when only a short length optical fibre of around 2 m is used. The dispersion will cause the pulses to have a pulse length of several picoseconds when they reach the scan head. The second effect is that the high peak power of the laser pulses can cause several non-linear optical effects to take place. These can include Raman signals from the glass, harmonic generation and self-phase modulation. All these processes have the effect of reducing the pulse energy and adding additional wavelengths to the light source, which leads to a loss of excitation power and unwanted light at the normal fluorescence wavelengths. A description of these effects can be found in a previous book chapter by the author (Girkin 2008).

The effect of pulse lengthening can be compensated for using methods known as "dispersion compensation". In reality all these methods use optics, typically before the scan head, to reshape the optical spectrum. In normal glass materials the blue light is slowed more than the red light as stated above. However, using prisms, diffraction gratings or special optical coatings on mirrors it is possible to arrange for the blue portion of the spectrum to be at the start of the optical pulse (Figure 8.6). The optics are set up such that the red portion of the

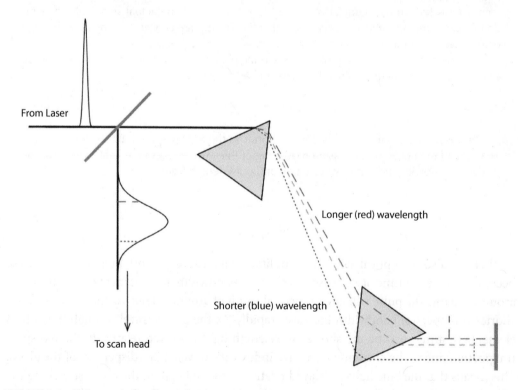

From Laser

Longer (red) wavelength

Shorter (blue) wavelength

To scan head

FIGURE 8.6 Prism pair for dispersion compensation.

spectrum travels a longer distance and thus arrives after the blue portion of the laser pulse at the scanning optics. The delay in the red wavelengths arriving can be adjusted such that at the focus of the microscope objective the red and blue portions of the beam arrive together to produce the required femtosecond pulses. Some commercial lasers now incorporate such dispersion compensation systems though they are not in routine use. The ability to compensate for this loss of pulse length due to dispersion can be required when imaging very deeply within a sample when one really needs everything to be working optimally.

In summary, unless one is trying to image very deeply, pulses of around 100–200 fs with a repetition rate of around 80 MHz work very well for most multiphoton applications. Although dispersion compensation can help to improve the signal to noise and depth of imaging this is not really required for routine multiphoton imaging.

8.2.3 Average Power

Multiphoton imaging typically requires about 500 times more average power on the sample than in conventional confocal laser scanning microscopy in order to achieve a similar number of excitation events within the focal volume. This is assuming a typical Ti:Sapphire laser operating at 80 MHz with 100–200 fs pulses. The effect of the average power on the sample (apart from affecting the peak power) is to cause heating within the sample due to linear absorption. Below 850 nm heating is not normally a significant problem unless the sample is heavily pigmented. As the wavelength rises above 900 nm single photon water absorption can become an issue and may be the predominating factor in limiting the use of multiphoton imaging in live samples.

The spectral width of the absorption curves and the tunability of Ti:Sapphire lasers can, however, be used to an advantage in certain situations to aid the problems caused by the imaging on the biological function of the sample. If one just considered the absorption profile then eGFP would be excited at 950 nm. However, the performance of Ti:Sapphire lasers is beginning to decrease in this spectral region, with a reduction in output power and in older systems pulse stability. Typically, therefore, eGFP is excited around 920 nm; however, for certain deep imaging applications, when the power has to be increased, linear absorption and heating in water is a problem. By tuning the laser to around 880 nm the eGFP is still excited (though not as efficiently) but the water heating is reduced by a factor of three, leading to an overall signal increase, in this case, of 20%. The crucial lesson from here, yet again, is to produce an image with sufficiently good signal to noise and resolution to enable the required quantification, but with minimal light exposure and perturbation to the sample.

Typically for femtosecond sources the power on the sample is under a few mW for most normal imaging challenges. This figure is, however, significantly affected by the sample, the excitation wavelength and the numerical aperture of the objective. As one images deeper into the preparation it may be necessary to increase the power as scattering events start to degrade the focus of the incoming beam and scatter the emitted light away from the collection optics. This is one of the major features of deep imaging by multiphoton excitation where, without the linear absorption of the excitation light, the illumination can be increased to maintain the image quality.

8.2.4 Objective Lens Selection

In any epi-illumination system (where the excitation light and signal are collected through the same lens) the selection of the objective lens plays a crucial role. A few guidelines on lens selection are provided below for the less experienced multiphoton user. After a time one tends to develop a preference for a lens to try first when a new imaging challenge is presented.

Lens NA: The simple rule here is normally to select the highest NA possible within the other constraints discussed below. The higher the NA, the smaller the excitation spot size and hence the greater the resolution and the more intense the excitation leading to a larger signal. Higher NA lenses also have significantly greater collection efficiency, as discussed earlier, and thus these are two strong drivers for selecting a high NA lens.

Lens Magnification: This really determines the field of view as the resolution in a beam scanned system is mainly determined by the imaging spot size and hence the NA of the lens. Higher magnification lenses also tend to have shorter working distances. When using a lower magnification lens, one can zoom in using the scanning zoom on the microscope if a smaller field of view is required.

Working Distance: After the NA of the lens this is probably the most important parameter to consider. In general one has selected to use multiphoton imaging because of the depth at which the technique can still produce high quality images and thus a longer working distance is normally preferred. The working distance and NA, for a given lens type, are directly linked and thus one may need to trade off the requirement for the highest NA possible with the requirement to image more deeply. Oil objectives, with their higher NA, are normally limited to around 150–200 μm and therefore are not often used due to their lack of working distance.

Air vs Water vs Oil: The selection here is frequently determined by the experimental protocol being developed. Whereas for confocal microscopy the first choice may well be an oil objective this is much less likely in the multiphoton case. As there is a good chance that one will be imaging through a depth of tissue, frequently *in vivo*, then the air objective (even if more limited in NA) is frequently a good lens to consider. Lenses are now available with NAs of 0.75–0.8 with a working distance of over a millimeter. The alternative for in depth imaging is to use a water-dipping lens. Clearly this limits the sample to being mounted on an upright microscope (unless one wishes to make special water retaining collars, etc.) but it enables higher NA lenses to be used and also helps to minimize the refractive index mismatch of oil between the tissue and the air. The use of a water-dipping lens also means there is less light scattered back from the surface of the tissue.

Summary of Lens Selection: In order to maximize the collection efficiency one should select a lens with the highest numerical aperture possible to allow for imaging into the required depth of the sample. A summary of typical lenses is presented in Table 8.2.

TABLE 8.2   A Summary of Objective Lenses Suitable for Use in Multiphoton Microscopy. (The Exact NA, Magnification and Working Distance May Vary Depending on the Manufacturer but All of the Major Suppliers Have Lens Parameters Close to These Values.)

Lens	NA	WD	Lat Res	Cost	Sample Mounting
×60 PA Oil	1.4	165 μm	0.25 μm	$$$$	Image through coverslip
×60 PA Water	1.2	220 μm	0.29 μm	$$$$$$	Image through coverslip
×40 SF Oil	1.3	220 μm	0.27 μm	$$$	Image through coverslip
×100 Plan Oil	1.25	170 μm	0.28 μm	$	Image through coverslip
×20 PA Air	0.75	1 mm	0.46 μm	$$	Inverted or upright
×20 PF Air	0.5	2 mm	0.69 μm	$	Inverted or upright
×10 PA Air	0.45	4 mm	0.77 μm	$$	Inverted or upright
×60 FL Water Dip	1.0	2 mm	0.35 μm	$$$	Upright
×40 Plan Air	0.65	480 μm	0.53 μm	$	Inverted or upright

Key

NA	Numerical Aperture.
WD	Working Distance.
Lat Res	Lateral Resolution (800 nm excitation).
PLAN	corrected for 2 wavelengths in visible.
PF	Plan Fluor, as Plan plus designed for high visible light transmission.
PA	Plan Apochromat, as Plan plus corrected for 4 wavelengths in entire visible spectrum.
SF	Super Flour, designed for maximum UV transmission down to 340 nm and visible.
FLUOR	As SF, designed for UV and IR transmission.
Cost	Indication of relative cost.

This is a slight simplification as it ignores the infrared light transmission through the lenses and aberrations present, but the approximation provides a very practical starting point for lens selection in a multiphoton microscope. The author frequently starts looking at a new sample using an air lens with NA of 0.75 and a working distance of 1 mm. As this lens does not mean one has to add oil to the sample or initially consider adding water it provides an excellent place for the user to start prior to potentially optimizing the system for a specific sample preparation.

8.2.5 Detection

As mentioned at the start of the chapter the aim should always be to direct the light to the detector as soon as possible with the minimum number of optical components, hence the preference for correctly implemented non-descanned detectors. Multiple labelling is routinely possible with multiphoton excitation but care should be exercised for the best combinations of fluorophores. The detector should be selected to match the emission wavelength of the fluorophore and dichromatic filter to increase the overall collection and detection efficiency. Details on specific detectors can be found in Chapters 2 and 5. Care should also be taken with detectors to ensure that they are not being affected by stray room lights or the image being displayed on the monitor.

In a multiple channel detection system at least one detector should have a short wavelength Bi-alkali detector and this should be the one to which all light is sent if the system is operating with no wavelength selective optics (dichromatic filters). In the more standard three-channel system the best combination would seem to be two Bi-alkali and one S20 detector. The advent of semiconductor avalanche detectors is now changing the position slightly and it can be anticipated that silicon-based sensors are likely to be employed for wavelengths where their overall efficiency is higher than those provided by the best photomultipliers. As silicon detectors have a range of spectral responses depending on the semiconductor doping there is no "rule" that can be suggested as being the optimal solution for detector combinations.

As a final comment on the selection and operation of multiphoton detectors it is always worth considering inserting an extra near infrared blocking filter before the detector. Such dichromatic filters, that pass visible and reject near infrared light, generally have a blocking factor of 10,000 and a transmission of over 95% in the visible. Thus they remove very few signal photons, and yet they can help remove any residual near infrared light that may have passed through the earlier filters.

8.3 GOING FROM CONFOCAL TO MULTIPHOTON MICROSCOPY

Most people using a multiphoton microscope for the first time will already have had experience of using a confocal microscope, and although there are many similarities in the way that the systems operate there are a few differences and the following points are aimed at helping first time users as they move on to multiphoton microscopy.

1. The first point to remember is that inherently the ultimate resolution that can be achieved with a multiphoton microscope will be lower than that achieved with a confocal. This is due to the larger excitation spot size as a result of the use of longer wavelength sources. Thus if you are trying a multiphoton microscope for the first time with a thin fixed sample, it is possible to be disappointed with the initial results as they may look a little more blurred. However, it should always be remembered that the main reason for switching to a multiphoton system is to image at a depth where confocal imaging cannot go, or on a sample where the visible excitation light of a confocal can induce damage.

2. If possible it is worth imaging a new sample using either conventional epi-fluorescence or confocal excitation (with a large pinhole) before switching over to the multiphoton source. It is appreciated that this is not always possible but it enables one to at least begin to orientate the view within the sample. Multiphoton systems can only produce an optically sectioned image and this can make it difficult to determine where one is in the sample to begin with. Thus when one first moves over to using multiphoton imaging it is worth spending time looking at a sample that one really knows and understands. This will enable new users to begin to gain an appreciation of the differences in suddenly only being able to see a single optical section. One cannot open up the pinhole, as in the case of a confocal, to obtain thicker samples showing a greater number of features to help with orientation. In the single multiphoton excited section one will also only have fluorescent features showing which can add

to the initial confusion. The use of conventional one photon excitation to begin with also means that one is certain that the sample is fluorescent, and if nothing is seen with the multiphoton system then the problem is not with the sample.

3. If one does start with the confocal and then changes sources it is possible that the first images may be from a few microns deeper into the sample. This is due to the chromatic aberration within the objective lens, which can shift the effective focal point into the sample. With some oil objectives at around 950 nm this shift can be as much as 5–10 microns.

4. The gain on the detectors will generally be set higher than that used for visible excitation and it should not be a major worry starting with the gain set higher than for confocal imaging. Once something can be seen on the screen the gain can be turned down as one optimizes the system.

5. The great temptation is to start with a high excitation power but if possible this should be avoided. The power can always be increased but if one starts high and damages the sample then there is nowhere to go except to prepare another sample!

6. As mentioned above, one needs to determine the initial excitation wavelength. Table 8.1 provides a suggested starting wavelength for a few common fluorophores and publications also indicate the imaging wavelengths used previously. As a quick guide, compared to the one photon wavelength a starting point is to try using a wavelength that is perhaps 50 to 100 nm less than the doubled wavelength of the one photon excitation peak.

7. As one images more deeply it may be necessary to increase the laser power as the spot will become larger at depth and excitation light will be lost due to scattering. If the power is increased it is important to remember to decrease the power as one comes back to the surface or damage can ensue.

8. If the system is fitted with a transmission detector this can be useful in helping to navigate around the sample. The transmission image may not be fluorescent but instead the transmitted light through the system is re-imaged using the illumination condenser. Although this image will not be a high resolution image it can help to set a fluorescence only optical slice in context.

9. In a sample where hair may have been shaved off be careful to ensure that no hairs are lying on the surface that one is imaging through as they can absorb the near infrared light and lead to localized charring.

10. As the wavelength is altered the signal may change either because one is moving away from the peak of the laser's output, or because the objective lens transmission is falling. Quantifying the power emerging from the objective, using either a simple power meter or the device mentioned in Section 8.4.3, is useful to understand such issues.

As a final piece of advice, be patient. There is a learning process to go through in starting to use any new technique, and thus creating time, and a few samples to experiment with at first, is worth the effort in the long term. Some typical example images are given in Figure 8.7 and many more can be found in more recent publications. It should be noted that no post processing has been undertaken on these images except for selecting the most suitable look-up table for display in this book.

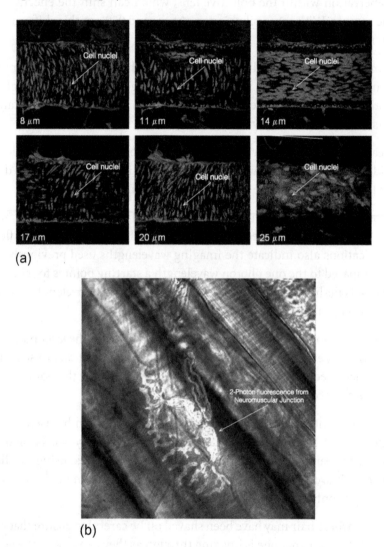

FIGURE 8.7 (a) Two photon image sections through an intact, living, blood vessel stained with DAPI selectively labeling nuclei. Starting with top left image and moving clockwise, sections show the upper smooth muscle cell layer, going through to the endothelial cell layer that lines the lumen of the vessel and back out through the muscle cells and then outer adventitial layer. The layers can be identified by the orientation of the nuclei. DAPI excited at 750 nm, 1 mW on sample, Endogenous fluorescence excited at 850 nm 1.5 mW on sample (sample prepared by Prof Alison Gurney), (b) Mouse neuromuscular junction (diaphragm) in white superimposed (1047 nm, 1 mW on sample) superimposed on a DIC image.

8.4 ADVANCED MULTIPHOTON MICROSCOPY

Due to the ability to image at depth, and in live samples, methods are rapidly evolving using multiphoton excitation. One area is the use of miniature optics in probes for imaging within live samples, frequently the brain. The second, the application of adjustable (adaptive) optics to compensate for aberrations induced by the sample, specifically when imaging at depth. At present there are no commercial systems readily available for this type of imaging. If this is an area of interest for the reader you are advised to read the latest literature on the subject, but the two short sections below provide some background on the directions that current research is moving in these two specialized fields.

8.4.1 Endoscopic Multiphoton Microscopy

A major role for multiphoton microscopy is in the *in vivo* imaging of activity within the brain, frequently using rodent models. The longer wavelength excitation means that imaging to a depth of a millimeter is possible and using three photon excitation depths of 2 mm can be achieved. However, there is a desire to image activity deeper than the top 2 mm of the brain and this has led to the development of miniature probes, which can move the focal point deeper into the sample.

The approach has been to re-image the focal point of a conventional microscope deeper using a set of miniature, rigid optics. Typically these probes are less than 1 mm in diameter and inserted into the sample. Although designs are possible using conventional optics most of the probes use graded refractive index (GRIN) lenses. These are lenses with flat surfaces and where the refractive power is provided by a refractive index that varies across the lens throughout its length. Technically this was originally achieved with a single, but long, optical element (Levene et al. 2009). Most designs now use a series of shorter lenses in order to minimize the dispersion and chromatic aberration.

As shown in Figure 8.8a, a conventional beam scanned multiphoton microscope is positioned above an endoscopic rod (Jung and Schnitzer 2003). The focal point of the imaging objective is above the GRIN rod optics mounted in the endoscope. The scanned imaging plane is then relayed through the GRIN optics into the sample. This produces a conventional multiphoton optical slice but from deep within the sample. The focus of the microscope objective can then be advanced to record further optical slices from greater depths. The advantage of this endoscopic imaging method is that all the beam scanning is undertaken outside the sample, meaning that the imaging endoscope can be smaller and the device can be made to fit onto a standard multiphoton microscope. The disadvantage is that the GRIN lens induces significant chromatic aberration, meaning that the light collection is optimal at a different focal plane to that where the excitation is present. It also means that the probe is rigid, limiting the movement of the sample.

An alternative approach is to place the scanning system close, or even inside, the sample. This has been achieved either by scanning a fibre inside the sample and relaying that scanned point onwards using a GRIN lens (Bird and Gu 2003; Figure 8b) or using a MEMS scanner mounted inside the sample connected to the outside world through an optical fibre and simple electrical connections (Engelbrecht et al. 2008; Le Harzic et al. 2008; Helmchen et al. 2001; Hoy et al. 2008; Figure 8c). These probes are inherently larger and

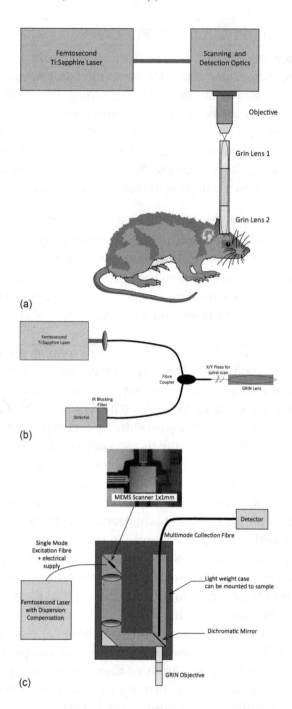

(a)

(b)

(c)

FIGURE 8.8 Three examples of endoscopic imaging using multiphoton excitation: (a) multiphoton microscopy using a miniature GRIN optics to relay the image into the sample through a small incision, (b) scanning the fibre inside the sample and imaging deeply using a GRIN lens, (c) A miniature scanning system suitable for mounting on an animal.

need electrical power to be delivered to the scan head. However, they are more flexible and hence the sample can be freely moving. These later versions are also being trialled for use in humans with the potential aim being for cancer diagnosis.

8.4.2 Adaptive Optics for Aberration Correction

As has been stressed throughout this chapter the advantage of multiphoton microscopy is to image at depth, frequently *in vivo*. However, as one images more deeply the excitation light has to pass through more of the sample and light can then be lost due to scattering. In addition to this loss of light at the focus due to scattering as the light travels through the areas of the sample with different refractive indices the wavefront is distorted leading to an aberrated focal spot. In non-linear microscopy this not only decreases the resolution but also significantly affects the local excitation intensity and hence the signal level. This is a similar challenge to that faced by ground-based optical telescopes where the light travelling through the earth's atmosphere distorts the image. To overcome these distortions adaptive optics have been developed in which the optical power of an element can be altered to compensate for the sample or atmospheric induced distortion (Booth 2007; Girkin, Poland and Wright 2009). Although systems incorporating such active aberration correction are not widely available commercially they are of great research interest and in use for those specifically looking to image deeply into the brain.

The basic configuration of a multiphoton microscope incorporating adaptive optics is shown in Figure 8.9. The light from the femtosecond laser is initially directed onto the scanning optics and then expanded and relayed onto the adaptive optic element. In a microscope the element is typically a continuous membrane mirror, with micro-actuators behind the mirror, capable of locally pushing or pulling the mirror surface to provide some local optical power. Mirrors typically having upwards of 37 actuators. Earlier systems tended to only have mirrors that were distorted by electrostatic attraction and thus could only pull the mirror. The light is then reflected off the mirror and directed onto the sample through the objective lens. In the configuration shown here (Marsh, Burns and Girkin 2003) the light passing to and from the adaptive optic mirror passes through a quarter-wave plate which on its second pass changes the polarization such that it is now reflected rather than transmitted by the polarizing beam splitter. The adaptive optic element can also be mounted before the scan head and the choice is basically one of convenience. In the system shown the light is then collected by a non-descanned detector.

The most important aspect in adaptive optics is to determine the best shape to apply to the mirror to remove the sample-induced aberrations. In astronomy, where the aberrations are highly dynamic due to atmospheric turbulence, a high power laser is often used to generate an artificial "guide star" but this is not normally practical within biological samples. Instead an image metric is used to assess the quality of the image as different mirror shapes are placed onto the mirror. Metrics include parameters such as contrast, brightness and image sharpness. The shapes on the mirror can either be carefully controlled (using so called Zernike polynomials of the light beam) or through more mathematical algorithms

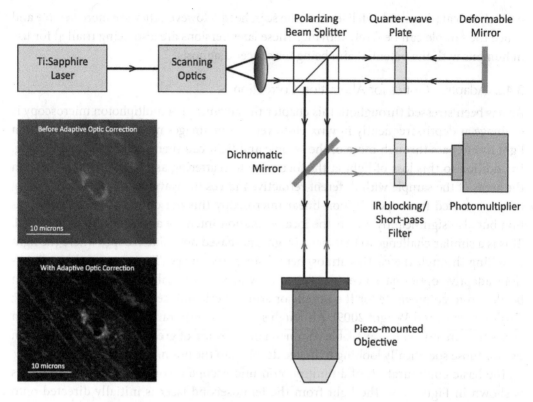

FIGURE 8.9 An optical system demonstrating the position of adaptive optics in a multiphoton imaging system. Beam relaying optics have been omitted for clarity. Inset includes images with and without the adaptive optics activated (Images from Marsh, Optics Express 2003, by permission of Optical Society of America).

including hill climbing and genetic algorithms to maximize the specific imaging metric (Lubeigt et al. 2010; Wright et al. 2005). The aberration correction can either be undertaken for every optical slice or determined for a series of slices, and the optimization does not need to take place for every image stack but predetermined mirror shapes can be used through the application of look-up tables.

Although not yet routine for deep multiphoton imaging, the method has enabled certain samples to be viewed at greater depth than was previously possible, and with higher resolution. It is also interesting to note that adaptive optics not only correct for sample induced aberrations but also for misalignment and distortions in the scanning and microscope optics. Thus, the added complexity of adaptive optics can help general multiphoton microscopy, and indeed all forms of both widefield and beam scanned imaging.

8.4.3 Measurement of Two Photon "Dose"

One of the major requirements in undertaking any form of quantitative biological or other scientific investigation is to ensure that the conditions when experiments are repeated are the same. When this requirement is placed on multiphoton imaging there are a large number of variables present which need to be measured. In respect to imaging, the optical system can be assumed to be the same but the laser parameters can change both from day to day and even during a single day. The critical requirements on the laser source are:

1. Wavelength

2. Average power

3. Pulse length

In the current computerized laser systems the wavelength is selected via the front panel and so this is being measured by the system. On older laser systems most users have a small scanning spectrometer which is used to set the wavelength and to ensure good mode-locking of the source. The average power can be measured using a standard power meter normally supplied with the laser. The pulse length is, however, significantly harder to measure.

For an accurate measurement an auto-correlator should be used as this gives a precise measurement of the actual pulse length. Such instruments are not easy to use and they are not suitable for measuring the pulse length at the sample (after the objective). A simple system is used by the author to monitor the two photon dose on the sample based upon non-linear absorption in a semiconductor device.

A GaAsP infrared blind photodiode (e.g. Hamamatsu G1116) can be mounted in a holder behind a focusing lens (12 mm diameter, focal length between 10 and 30 mm). A small portion of the incoming laser beam is sent into the detector before the beam enters the scan head but after any intensity controlling optics (neutral density filters, AOTFs or Pockels' cell) and miniature spectrometers, Figure 8.10. A 100 k ohm resistor is mounted across the diode and a 200 mV digital voltmeter connected across this output. When the Ti:Sapphire pulses are incident on the detector the fundamental infrared light gives no signal on the meter, but when the focal volume is matched to the depletion region of the diode then two photon excitation of electrons into the conduction band takes place and a signal can be detected. This voltage is directly related to the peak power in the same way as the two photon excitation fluorescent signal in the sample. As the pulse lengthens so the voltage will fall. Through the correct choice of resistor, or focusing lens, up to four decades of useful range can be found. This gives a quantified measure of the output from the laser related to the pulse length and with a measurement of the average one photon power the laser output is characterized.

A second diode can then have its outer "can" removed and be mounted, with a similar resistor and meter, after the objective, and this will then give a measurement of the two photon "dose" at the sample. By undertaking a simple calibration process changing the power and wavelength the dose on the sample stage can be linked to that measured by the first detector. The stage detector can then be removed and the user can now make a measurement of the incoming two photon dose and relate this to the actual sample dose.

In the author's multiphoton system such a device is mounted for every laser source, along with a small infrared sensitive photodiode, which has been calibrated to measure the average power. Conditions for each image can now be recorded and care taken that for live cell studies the threshold damage power for the biological system is not exceeded. Such a device is low cost but enables quantifiable imaging to be undertaken.

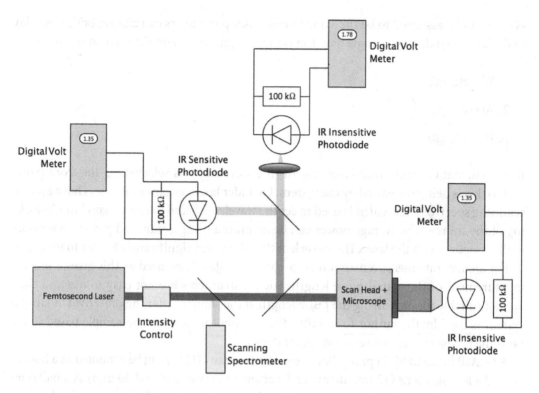

FIGURE 8.10 Two photon dose meter.

REFERENCES

Bird, D. and M. Gu. 2003. "Two-Photon Fluorescence Endoscopy with a Micro-Optic Scanning Head". *Optics Letters* 28(17): 1552–4. www.ncbi.nlm.nih.gov/pubmed/12956376.

Booth, M. J. 2007. "Adaptive Optics in Microscopy". *Philosophical Transactions. Series A, Mathematical, Physical, and Engineering Sciences* 365(1861): 2829–43. www.ncbi.nlm.nih.gov/pubmed/17855218.

Denk, W., J. H. Strickler and W. W. Webb. 1990. "Two-Photon Laser Scanning Fluorescence Microscopy". *Science (New York)* 248(4951): 73–6. www.ncbi.nlm.nih.gov/pubmed/2321027.

Engelbrecht, C. J., R. S. Johnston, Eric J. Seibel and Fritjof Helmchen. 2008. "Ultra-Compact Fiber-Optic Two-Photon Microscope for Functional Fluorescence Imaging in Vivo". *Optics Express* 16(8): 5556–64.

Girkin, J. M. 2008. "Laser Sources for Non-Linear Microscopy". In *Handbook of Biomedical Nonlinear Optical Microscopy*, eds B. R. Masters and Peter T. C. So. Oxford University Press.

Girkin, J. M. and G. McConnell. 2005. "Advances in Laser Sources for Confocal and Multiphoton Microscopy". *Microscopy Research and Technique* 67(1): 8–14. www.ncbi.nlm.nih.gov/pubmed/16025485.

Girkin, J. M., S. Poland and A. J. Wright. 2009. "Adaptive Optics for Deeper Imaging of Biological Samples". *Current Opinion in Biotechnology* 20(1): 106–10. www.ncbi.nlm.nih.gov/pubmed/19272766 (July 24, 2011).

Goeppert-Mayer, M. 1931. "Ueber Elementarake mit Zwei Quantenspruengen". *Annalen der Physik* 9: 273–83.

Hanninen, P. E., E. Soni and S. W. Hell. 1994. "Continuous Wave Excitation Two-Photon Fluorescence Microscopy". *Journal of Microscopy* 176: 222–5.

Le Harzic, R., M. Weingel, I. Riemann, K. König and B. Messerschmidt. 2008. "Nonlinear Optical Endoscope Based on a Compact Two Axes Piezo Scanner and a Miniature Objective Lens". *Optics Express* 16(25): 20588–96. www.ncbi.nlm.nih.gov/pubmed/19065197.

Helmchen, F., M. S. Fee, D. W. Tank and W. Denk. 2001. "A Miniature Head-Mounted Two-Photon Microscope. High-Resolution Brain Imaging in Freely Moving Animals". *Neuron* 31(6): 903–12. www.ncbi.nlm.nih.gov/pubmed/11580892.

Hoy, C. L., N. J. Durr, P. Chen, W. Piyawattanametha, H. Ra, O. Solgaard and A. Ben-Yakar. 2008. "Miniaturized Probe for Femtosecond Laser Microsurgery and Two-Photon Imaging". *Optics Express* 16(13): 9996–10005. www.ncbi.nlm.nih.gov/pubmed/18575570.

Jung, J. C. and M. J. Schnitzer. 2003. "Multiphoton Endoscopy". *Optics Letters* 28(11): 902–4. www.ncbi.nlm.nih.gov/pubmed/12816240.

Levene, M. J., D. A. Dombeck, K. A. Kasischke, R. P. Molloy and W. W. Webb. 2004. "In Vivo Multiphoton Microscopy of Deep Brain Tissue". *J Neurophysiol* 91(December 2003): 1908–12.

Lubeigt, W., S. P. Poland, G. J. Valentine, A. J. Wright, J. M. Girkin and D. Burns. 2010. "Search-Based Active Optic Systems for Aberration Correction in Time-Independent Applications". *Applied Optics* 49(3): 307. www.osapublishing.org/abstract.cfm?URI=ao-49-3-307.

Marsh, P., D. Burns and J. Girkin. 2003. "Practical Implementation of Adaptive Optics in Multiphoton Microscopy". *Optics Express* 11(10): 1123–30. www.ncbi.nlm.nih.gov/pubmed/19465977.

Wilson, T. and C. J. R. Sheppard. 1984. "Non-Linear Microscopy". In *Theory and Practice of Scanning Optical Microscopy* Wilson and Shepphard (Eds.). Boston: Academic Press.

Wright, A. J., D. Burns, B. A. Patterson, S. P. Poland, G. J. Valetine and J. M. Girkin. 2005. "Exploration of the Optimisation Algorithms Used in the Implementation of Adaptive Optics in Confocal and Multiphoton Microscopy". *Microscopy Research and Technique* 67(1): 36–44. www.ncbi.nlm.nih.gov/pubmed/16025475.

CHAPTER 9

Harmonic Microscopy

H ARMONIC MICROSCOPY IS BASED upon a non-linear effect that takes place within certain materials and has a totally different physical origin to fluorescence. The effect of the harmonic generation of light is not due to the absorption and re-emission of light but a result of intense light fields causing changes in a molecule's local electric field. As there is no absorption of light the process takes place without the transfer of energy to the sample (assuming there are no other molecules that can absorb the incoming light). The process can take place in molecules that are already present in the sample and thus no exogenous compounds need be added, nor is any genetic manipulation of the sample required such that it produces the fluorescent molecules itself. This would appear to make harmonic microscopy an excellent contrast mechanism with minimal perturbation of the sample. However, there is one significant drawback. As will be described below, although the harmonic generation can take place in nearly all molecules, in most cases the light produced interferes with itself and so there is no emission of detectable photons. In order to observe harmonic generation in microscopy, the molecules need to be aligned such that the interference process is constructive rather than destructive.

The effect was known of for many years but as with non-linear fluorescence the phenomenon was only seen in the optical regime with the invention of the laser. It may interest some readers to know that in the first publication on harmonic generation (Franken et al. 1961) the editorial team removed the spectral signature of the harmonic generation from the spectral photograph. The editor thought it was dust on the negative!

We will first examine the physical basis for harmonic generation before considering how this contrast mechanism can be utilized for microscopy and the optimal instrumentation for practical imaging in this manner. As well as some examples with scientific importance a few very simple samples will be suggested to enable readers, with suitable equipment, to observe the process for themselves. We will also present methods by which a user can determine if some of the signal being detected in a multiphoton microscope is from harmonic generation rather than non-linearly excited fluorescence. As the process is non-linear with the excitation intensity, many of the comments made in Chapter 8 on multiphoton fluorescence microscopy will be shown to apply to harmonic imaging as well.

9.1 PHYSICAL BASIS FOR HARMONIC GENERATION

In optical harmonic generation, light enters the sample and the output is detected at exactly half, or a third, of the input wavelength (though in non-microscopy applications even higher order processes are possible). However, the physical mechanism behind the effect is not just two photons coming together to produce a single photon at twice the energy, or half the wavelength, though that is the apparent effect. The formation of harmonics (frequencies that are an exact multiple of a fundamental frequency) is a phenomenon of so called *"classical physics"* and is a result that can be obtained by solving Maxwell's laws of electromagnetism. Indeed the effect of harmonic generation can occur for any wave (electromagnetic or sound) that has sufficient intensity when it is incident on a material. Harmonics (multiples of the fundamental wavelength) can be found, for example, in stringed musical instruments, where as well as vibrating at the main wavelength the string in a violin will also oscillate at multiples of that frequency. In this case it helps to produce the "tone" of an instrument. The process for light waves is the basis of all current green laser points where the green light at 532 nm is produced by focusing the fundamental laser light at 1064 nm (in the infrared) onto a crystal.

Light, as explained in Chapter 2, is an electromagnetic wave and as such when it impinges on a molecule it changes the electric field of that molecule. This effect is very small. The light wave causes the molecule to develop one end which is slightly more positive while the other end becomes slightly negative. This *"polarization"* of the molecule is known as a dipole, which adds to the inherent dipole (electron distribution) of the molecule without the external electric field. The total of all such additional dipoles for a molecule consists of a series of terms that are driven to oscillate by the incoming light wave. The first term responds to the fundamental electric field, but the second term responds to the square of the electric field (one part causing the dipole change, the other the normal propagation of the light). Although small in magnitude these additional dipoles can become significant if the electric field is high enough. At very high light intensities these higher order interactions are driven by the light and oscillate at twice the frequency of incoming light. This leads to a second electromagnetic wave at twice the frequency of the original light as illustrated in Figure 9.1.

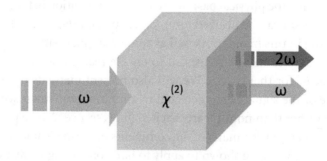

FIGURE 9.1   Second harmonic generation in a material. The direction of the arrows also indicates the direction of propagation of the light.

If one considers light as a wave hitting a molecule the lightwave's electric field will cause changes in the charge distribution within the molecule, inducing an electric dipole moment for that molecule. Under normal circumstances this effect is very small and the size of the dipole is linear with the electric field. The dipole moment per unit volume of the material (intrinsic and induced by the electromagnetic wave) is known as the polarization. The total polarization (P_{total}) can then be considered as the intrinsic value with a subsequent series of additional changes that depend on the incoming electric field (E). Mathematically this is expressed in a simplified form as

$$P_{total} = P_1^0 + \chi^{(1)}E + \chi^{(2)}E^2 + \chi^{(3)}E^3 + \ldots$$

where the term P_1^0 is the permanent polarization, $P^1 = \chi^{(1)}E$ has the change in polarization being linear with the applied electric field, $P^2 = \chi^{(2)}E^2$ (the second order polarization) has the change varying with the square of the electric field, and so on. Full details can be found in any textbook on non-linear optics such as Boyd (2008). The light wave (electric field) is just propagated by the P^1 term but the second term causes both a linear continuation of the light field and crucially, a wave in which the electric field has exactly twice the incoming frequency. At normal intensities of light this effect is very small but in high electric fields (such as those seen in high power lasers) this harmonic wave can become large enough to be detected and even seen. The simple effect is illustrated in Figure 9.1 with the χ^2 coupling with the incoming wave of frequency ω to produce an output containing light at two frequencies ω and 2ω.

The diagram also illustrates that the wave produced travels in the same direction as the incoming light. Thus the emission is in the forward direction in an imaging system and this can be important in the detection method selected, as discussed towards the end of Section 9.2.

There is one crucial aspect to this light production and driving of the molecular dipoles that is vital in determining if any second harmonic light will be detected. The process described above is coherent (in phase) with the incoming waves of light and thus all light produced by this harmonic process is coherent. In most normal situations this means that the waves produced by one molecule will destructively interfere with the waves from the molecule next door. In molecular arrangements with a non-centrosymmetric structure the light fields produced are such that they do not cancel out and the second harmonic is visible. This effect continues for the third harmonic and further up the frequency chain but with decreasing amplitude. One application of harmonic imaging is for viewing single molecule surfaces where all the molecules can be aligned in the same direction due to forces such as surface charge and surface tension. The molecules will then all oscillate in phase in response to the incoming light, producing constructive interference.

This process can therefore be summarized simply. At a base level one can think of the molecules as vibrating in response to a wave. If the amplitude of the wave becomes very large then harmonics can be produced at twice, three times, and so on up the series, the frequency of the fundamental wave. In many cases the individual molecules will be arranged in such a way that these harmonics all cancel out, but with the correct pattern they will all be in phase and the harmonic will be seen.

9.2 PRACTICAL HARMONIC MICROSCOPY

In order to use this physical process for practical microscopy applications there are several important features to be considered. The first, as with multiphoton imaging, is that the process depends non-linearly on the intensity of the light source. In the case of second harmonic imaging (SHG imaging) the intensity squared, for third harmonic imaging (THG imaging) the intensity cubed. Thus the process will inherently be limited to a small excitation volume as illustrated in Figure 8.1. The scale of this excitation volume is determined by the diffraction limit (set by the wavelength and numerical aperture of the lens, as described in Chapter 2) and the square of the electric field. However, in some cases, such as at interfacial surfaces, one has optical sectioning that is determined by the thickness of the layer as molecules below this layer may not be aligned. This can provide a resolution of tens of nanometers, though perhaps this should be considered as single molecule imaging rather than super-resolution microscopy.

Due to this non-linear response to the excitation light, the most commonly used sources for harmonic imaging are the same femtosecond lasers used for multiphoton microscopy and for the same reasons. The longer wavelengths in the near infrared are generally not absorbed, they are scattered less and can thus penetrate more deeply into the sample. They have the required high peak power while maintaining a low average power that does not heat the sample.

The second major consideration is related to the wavelength of the emission light. The light that is detected will be at exactly twice (SHG) or three times (THG) the excitation source. This has two consequences. The first is that the emission wavelength can easily be separated from the source light as it has a very narrow spectral width at a known wavelength. Second, through choice of the correct fluorophores, harmonic imaging can be carried out alongside multiphoton fluorescence microscopy. With the correct filters present the fluorescence can be detected in one channel, and the harmonic signal in a second. For harmonic excitation the requirement to tune the source to a specific absorption wavelength is not necessary as the harmonic process will take place at all wavelengths in the visible and near infrared. Thus the laser can be tuned to the fluorophore peak (say around 950 nm for green fluorescent protein) and the fluorescence will appear between 500 and 570 nm and the harmonic signal at 425 nm. Clearly though one would not want the emission light to be strongly absorbed by a compound in the sample. The flexibility in harmonic imaging is such that one might occasionally tune the excitation wavelength so the harmonic avoids any risk of absorption within the sample.

As well as the light emerging at exactly half the incoming wavelength, the harmonic light is also emitted instantly. There is no time lag between the excitation light and the signal. This has two important features when used in conjunction with fluorescence. The first is that the harmonic signal can be separated from the fluorescence using a gated detector, as described in Chapter 6. This is clearly a positive feature but the rapid response can also be a disadvantage. If one is trying to measure a fluorescence lifetime and there is harmonic light reaching the lifetime detector this will frequently mean that the detector is triggered before any fluorescent photons have been released. In this case the harmonic light may need to be removed using a dichromatic mirror, or the detector activated around 0.5 ns after the excitation pulse and hence after the harmonic signal would have reached the detector.

The other crucial factor in the practical use of harmonic microscopy is the direction of the light emission. As mentioned above and shown in Figure 9.1 the harmonic light produced by the process is in the same direction as the incoming light. This is significantly different from fluorescence emission, which is emitted in all directions. Thus in a fluorescence microscope at least half the light comes directly back towards the microscope and detector. In the case of harmonic imaging the light naturally travels away from the detector. The positive feature here is that one is normally imaging in scattering samples and hence some of the harmonic light will be scattered back through the normal epi-illumination route.

There are two ways in which the signal collection can be improved, however. Figure 9.2 shows the optical configuration of a conventional harmonic microscope, which is clearly very similar to the multiphoton system described in Chapter 8. However, through the use of the conventional illumination optics the signal can be significantly improved. Inset "a"

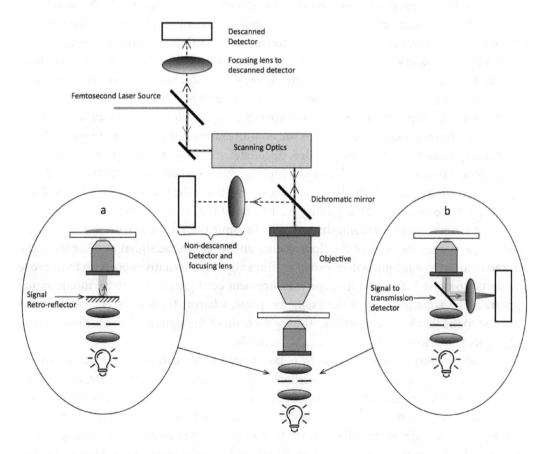

FIGURE 9.2 Harmonic microscope configuration demonstrating two optional improved detector options, a) retro-reflection using the condenser lens, b) transmission detector after the condenser lens.

indicates one way in which this can be achieved. Here the optics of the illumination system have been correctly adjusted for Köhler illumination and then a plan mirror placed at the focal point of the condensing optics. This mirror will then retro-reflect any transmitted light back through the system and into the epi-illumination optics. Clearly this will also send back the high intensity illumination light but this can easily be removed using a high performance dichromatic filter with a narrow pass band, which will only transmit the harmonic light. Even placing the sample on a mirror in place of a conventional microscope slide can also help to increase the harmonic signal.

An alternative enhancement method is shown in inset "b". Here a switchable mirror has been placed in the standard widefield illumination arm. After the sample had been set up in the normal way, again ensuring Köhler illumination, the mirror can be switched in so that any transmitted light is sent into a non-descanned transmission detector. Again a high performance filter is needed in front of the detector, or even two in some instances. Using this method the author has seen an order of magnitude increase in the detected harmonic signal in certain samples and thus the effort required to implement one of these transmission collection systems can be very worthwhile. It is also interesting to note that half of the conventional fluorescence light also passes in this direction and again, these same methods can be employed to increase the signal for conventional multiphoton microscopy.

All other considerations for efficient harmonic microscopy are the same as those described in Chapter 8 where multiphoton fluorescence microscopy was discussed. It is worthwhile noting that harmonic generation can actually be a real problem in some fluorescence imaging experiments. In certain experimental situations the harmonic signal can provide significant background signal in some faintly fluorescent emitting samples. As will be discussed in some of the examples described later, collagen is a structure that is highly amenable to harmonic imaging. Due to its high harmonic generation potential, if collagen is present within the sample the harmonic signal can mask the fluorescence. This problem can normally be overcome by adjusting the tuning of the laser such that any filters present reject the harmonic wavelengths but still transmit the fluorescence. This may mean moving away from the peak of the fluorescence emission but the signal to noise from the background harmonic light will improve significantly. An alternative approach to improve the signal to noise is to use a lifetime measurement configuration. The harmonic signal appears at the same instance as the excitation pulse, whereas the fluorescence will appear later due to the fluorescence lifetime. Thus separation of the signals can be achieved using either gated detectors, or photon counting methods.

There are several simple and easy to obtain samples that can be used to demonstrate harmonic imaging, and provide an excellent way to ensure that a non-linear microscope has the correct filters and detectors present to record harmonic images. Starch is a molecule that exhibits a strong second harmonic signal and is readily available in rice grains or even a potato. In order to initially confirm that a microscope can record harmonic images one of the two suggested samples can be placed under the microscope and the system set to scan. In the case of the potato the highly scattering nature means that the signal will be sent back towards the objective and will thus be efficiently collected by the conventional epi-illumination system present in non-linear microscopes. The second harmonic that is generated in these simple examples is likely to be visible to the eye. By using a pair of near infrared safety glasses, that are transparent in the visible, if the femtosecond laser is tuned to around 900 nm blue light should be visible on the sample at the focus of the objective. In most standard commercial systems this light should be detected on the blue detector (normally used when 405 nm light is being used to excite fluorescence in the confocal mode). If this is not the case then the laser wavelength should be adjusted, such that light at half the wavelength passes through the filter onto the detector.

As a rough guide slightly higher powers than might be used for conventional multiphoton fluorescence imaging may be required. Generally the high power is not a problem as the sample is unlikely to have chromophores present that will absorb the light directly, thus minimizing the risk of heating in the sample. However, as with all imaging, the power should be kept as low as possible to produce a high enough signal to noise image. The main reason in the case of harmonic imaging is to minimize the risk of the generation of unwanted, and potentially harmful, ultraviolet radiation through third harmonic generation. Optimization of the image follows the same guidelines presented on multiphoton microscopy, though the early comments about the direction of the harmonic signal should be considered.

9.3 APPLICATIONS OF HARMONIC MICROSCOPY

For most biological samples the materials are aligned so that they do not have second harmonic terms, or such that any harmonics that are produced interfere destructively. The first reference to the use of harmonic generation in microscopy was in 1977, with the first system that might have the sensitivity required for biological imaging demonstrated in the following year (Sheppard et al. 1977; Sheppard and Kompfner 1978). Roth and Freund then produced the first biological images in 1981 which was followed shortly afterwards with the use of the method to provide insight into a real biological question by imaging the structure of a rat tail (Roth and Freund 1981, 1982). The method has since gained widespread acceptance as a method that enables imaging without the need for exogenous compounds. As is the case for all non-linear imaging the use of the technique by multiple groups around the world was delayed until laser sources became significantly easier to use and the emphasis moved over to answering scientific questions and not one of just keeping the laser and optics aligned.

Third harmonic imaging, until recently, has been used much less frequently. The main use has been in the imaging of lipids (Barad et al. 1997; Supatto et al. 2006) though Raman-based methods described in Chapter 10 have generally superseded harmonic imaging in this case. Another interesting development, and practical application, was for imaging cell membranes though in this case a dye was added to label the lipids. Rather than excite the dye molecule by tuning to a two photon absorption band and detecting the subsequent fluorescence, the authors used the dye as a source of harmonic generation (Campagnola et al. 1999). This was achieved as the dye was in the cell membrane and thus in a very thin layer such that the molecules were generally aligned in one direction. This meant that there was no destructive interference of the harmonic light. Indeed this paper took the work one stage further as the dye was voltage sensitive, thus as the potential within the cell changed, fewer, or an increasing number, of the molecules aligned themselves within the membrane and thus the SHG signal provided a readout of the membrane potential. This illustrates one example of when harmonic imaging can have some advantages over more conventional fluorescence base methods. Here information was provided on the local molecular fluctuations.

The method has been applied to *in vivo* imaging not only in the cornea but, potentially more interestingly, as a method of tumour detection (Guo et al. 1999). This demonstrated that SHG was a viable tool for *in vivo* use where one does not need to add fluorophores. The advance in the method can be seen in work published more recently where human adipocytes have been imaged directly using third harmonic microscopy (Tsai et al. 2013). Both of these papers illustrate the major potential in harmonic imaging for *in vivo* human use as no exogenous compounds were introduced. The method can also be combined with more conventional two photon fluorescence where the SHG provides the information on the structure of the sample and the fluorophore (yellow fluorescent protein in the example in Figure 9.3) a specific biological tissue feature. In this case with the laser tuned to 950 nm (close to the two photon peak for YFP) the collagen and muscle fibres emitted at the second harmonic of 475 nm (well away from the YFP emission peak around 560 nm).

The main applications of harmonic imaging have now emerged as being the imaging of collagen structures within *in vivo* samples. Here the fact that a fluorophore does not need to be added means that the structure of the tissue can be seen while fluorescence is reserved for labelling of more specific features. If there are no chromophores present the harmonic method further reduces the risk of thermal damage to the sample as there is

Blue/Cyan = SHG (475 nm) – collagen and muscle fibres
Yellow = pericytes expressing YFP (exc = 950 nm)

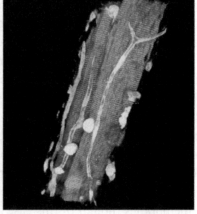

FIGURE 9.3 A combined SHG and multiphoton image of yellow fluorescent protein illustrating the advantage of a combination imaging method. The SHG provides the biological context for the structure and position of the YFP expressing pericytes. (Image courtesy of Dr Paul Thompson, Henry Wellcome Laboratory for Cell Imaging, University of East Anglia, in Biological and Medical Photonics, Spectroscopy and Microscopy edited by David Andrews 2015. With Permission from John Willey & Sons.)

little in tissue, except water at longer wavelengths, to absorb the laser radiation. The narrow band emission, and the ability to tune away from fluorescence if required, adds a further degree of flexibility in imaging samples. The other major application is in observing vesicle membranes where the very thin nature of the membrane, as described above, means that molecules can be aligned to enable constructive, rather than destructive interference.

REFERENCES

Barad, Y., H. Eisenberg, M. Horowitz and Y. Silberberg. 1997. "Nonlinear Scanning Laser Microscopy by Third Harmonic Generation". *Applied Physics Letters* 70(8): 922.

Boyd, R. W. 2008. *Nonlinear Optics*. 3rd edn. Elsevier.

Campagnola, P. J., M. D. Wei, A. Lewis and L. M. Loew. 1999. "High-Resolution Nonlinear Optical Imaging of Live Cells by Second Harmonic Generation". *Biophysical Journal* 77(6): 3341–49.

Débarre, D.,W. Supatto, A. M. Pena, A. Fabre, T. Tordjmann, L. Combettes, M. C. Schanne-Klein and E. Beaurepaire. 2006. "Imaging Lipid Bodies in Cells and Tissues Using Third-Harmonic Generation Microscopy". *Nature Methods* 3(1): 47–53.

Franken, P., A. Hill, C. Peters and G. Weinreich. 1961. "Generation of Optical Harmonics". *Physical Review Letters* 7(1): 118–20.

Guo, Y., H. E. Savage, F. Liu, S. P. Schantz. P. P. Ho and R. R. Alfano. 1999. "Subsurface Tumor Progression Investigated by Noninvasive Optical Second Harmonic Tomography". *Proceedings of the National Academy of Sciences of the United States of America* 96(19): 10854–6.

Roth, S. and I. Freund. 1981. "No Optical Second Harmonic Scattering in Rat-Tail Tendon". *Biopolymers* 20: 1271–1981.

Roth, S. and I. Freund. 1982. "Second Harmonic Generation and Orientational Order in Connective Tissue: A Mosaic Model for Fibril Orientational Ordering in Rat-Tail Tendon". *Journal of Applied Crystallography* 15: 72–8.

Sheppard, C. J. R. and R. Kompfner. 1978. "Resonant Scanning Optical Microscope". *Applied Optics* 17(18): 2879–82.

Sheppard, C. J. R., J. N. Gannaway, R. Kompfner and D. Walsh. 1977. "The Scanning Harmonic Optical Microscope". *IEEE Transactions in Quantum Electronics* 13: 100D.

Tsai, C-K., T-D. Wang, J-W. Lin, R-B. Hsu, L-Z. Guo, S-T. Chen and T-M. Liu. 2013. "Virtual Optical Biopsy of Human Adipocytes with Third Harmonic Generation Microscopy". *Biomedical Optics Express* 4(1): 178–86.

CHAPTER 10

Raman Microscopy

T HE USE OF THE Raman effect as a contrast mechanism in optical microscopy is growing through the imaging of indigenous Raman active materials and also via the addition of Raman labels: frequently deuterated versions of compounds already present in the sample. Although the effect was first recorded in 1928, and its basis explained through quantum mechanics shortly afterwards, it was a phenomenon waiting for technology to enable its exploitation. A Raman spectrum is frequently described as a molecular fingerprint, with each chemical bond producing spectral lines. Each spectral line can be associated with a vibrational mode of the molecule and specific chemical bonds. However, the Raman effect is very weak and when multiple molecules are present the spectra become highly complex. The advent of the laser meant that a source with very high spectral brightness (high power but with a very narrow spectral linewidth) was available and therefore interest in the method increased. In the last twenty years this interest has grown further as detectors have been improved alongside efficient computer algorithms for interpretation of what can be highly complex spectra. In parallel with the use of Raman spectroscopy for chemical analysis there has been an increased use of the Raman effect to obtain specific contrast in microscopy images. The effect is used in linear, beam scanned confocal Raman microscopy and more recently using complex non-linear excitation of the sample in methods such as Coherent Anti-Stokes Raman Scattering (CARS).

The Raman effect was first predicted in 1923 and takes place when light is inelastically scattered by a molecule. As the photon/molecule collision is inelastic there is an exchange of energy resulting in the photon emerging from the process with a changed wavelength, normally to lower energy and thus longer wavelengths. The first observation of inelastic light scattering was made in 1928 by C. V. Raman and K. S. Krishnan and led to the award of the Nobel Prize in Physics in 1930. Sir Chandrasekhara Venkata Raman, an Indian, was the first Asian and non-white to receive a scientific Nobel Prize and the effect was rapidly named after him. In their original work Raman and Krishnan used sunlight and a narrow band filter to produce "monochromatic light". This light was then passed through a sample and it was noted that a small amount of the light had changed wavelength as it passed through a filter that blocked the original monochromatic wavelength. Shortly afterwards mercury arc lamps and improved spectrometers were used to produce tables with spectral lines from a wide range of molecules. As only around one photon in 10 million is Raman scattered (Raman scattering probability $\sim 1 \times 10^{-7}$) the technology at the time was not sensitive enough for widespread practical application of the method. Instead near infrared spectroscopy became the standard analytical tool used to help determine molecular composition. The laser and more recently CCD detectors have radically transformed this position.

The chapter will explain the physical basis behind the Raman effect before presenting the practical methods now used to employ Raman scattering as a molecularly specific contrast mechanism in a wide range of samples. The chapter will then look at the development and application of non-linear Raman-based methods such as CARS. The basic configurations of Raman microscopes will only be described briefly, as at their core they are confocal and multiphoton instruments, but some practical guidelines will be given to maximize the performance of such instruments. As Raman microscopy is currently a rapidly evolving field the latest practical advances in the method will be highlighted. Several applications of Raman microscopy will also be presented to enable the reader to better appreciate when Raman microscopy may be of practical benefit.

10.1 PHYSICAL BASIS OF THE RAMAN EFFECT

The Raman effect is a light scattering phenomenon in which the electromagnetic field associated with light interacts with the electric field of the molecule, and in particular with the localized electric field around each molecular bond. The effects of light interacting with a molecule can be seen in Figure 10.1. When light interacts with a molecule most of the

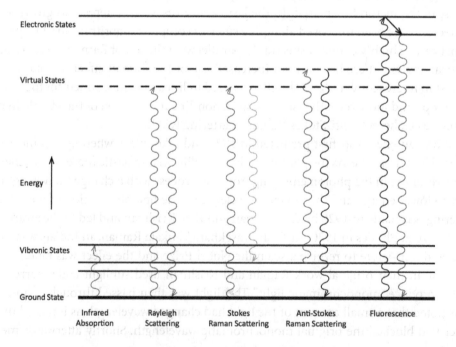

FIGURE 10.1   Light interaction with molecules showing Rayleigh scattering, Stokes shifted Raman scattering, anti-Stokes shifted Raman scattering and fluorescence.

time it will be elastically scattered, with the result that the emerging light has the same wavelength as the incoming beam and there is no exchange of energy with the sample. This type of inelastic scattering is known as Rayleigh scattering. If the light is in the near, or mid, infrared there is a possibility that the photon will be absorbed to alter the vibrational levels of the molecule, leading to a transfer of energy to the molecule, in effect heating the sample. If the photon has sufficient energy it can also be absorbed causing an electron to be excited to a higher electron energy level, leading to fluorescence. This is not a scattering event as was discussed in Chapters 2 and 6. However, around 1 photon in 10^7 can be scattered by the molecule and emerge from this encounter with a change in wavelength.

In classical terms the electric field of the light wave deforms the electric field of the molecule, causing the latter to oscillate at the same frequency as the light wave. The molecular oscillating field then acts as a dipole oscillator emitting light at the same frequency as the light field when there are no Raman active vibrational modes. An alternative is when the molecule is Raman active, part of the electromagnetic wave is used to excite a vibrational mode in the molecule and the resulting molecular dipole oscillates at a lower frequency, emitting light at a slightly longer wavelength than the incoming light. This emitted Raman light is known as the Stokes wavelength as it emerges at a longer wavelength. The third option is that the molecule is already in an excited vibrational state and when the light field interacts with the molecular field the energy from the vibrational mode and the light field is combined such that the emitted light is at a higher frequency (shorter wavelength) than the original photon. This produces anti-Stokes wavelengths. This latter effect has an even lower probability of occurring, as the molecule must already be in an excited vibrational state.

In a simple classical electromagnetism model the dipole moment, P, induced in a molecule by an external electric field, E, is proportional to the field via a proportionality constant $\chi^{(1)}$, known as the polarizability of the molecule:

$$P = \chi^{(1)}E \tag{10.1}$$

This polarizability term is a measure of how susceptible the electron cloud around the molecule is to being distorted. For asymmetric molecules (e.g. O–H) this additional effect is small, but for symmetrical molecules or chemical bonds (e.g. C=C) the effect is more significant. It is the induced dipole caused by this polarizability which emits, or scatters, the incoming light. Raman scattering then occurs when a molecular vibration changes the polarizability. This change is described by $\dfrac{\delta\chi^{(1)}}{\delta Q}$ where Q is the normal coordinate of the vibration. The selection rule for a Raman active vibration is that there has to be a change in the polarizability during the vibration or

$$\frac{\delta\chi^{(1)}}{\delta Q} \neq 0. \tag{10.2}$$

The scattering intensity is then proportional to the square of the induced dipole moment

$$I \propto \left(\frac{\delta\chi^{(1)}}{\delta Q}\right)^2. \tag{10.3}$$

The Raman effect can also be considered, and perhaps more easily appreciated, using a quantum mechanical approach (see Figure 10.1). The incoming photon excites the sample into a virtual energy state from which a photon is subsequently emitted with the molecule returning to its ground state that explains Rayleigh scattering. In the Stokes situation the molecule does not return to its ground state but to a vibrational state that is at a slightly higher energy, leading to the emitted photon having a slightly lower energy, or longer wavelength. The final alternative is for the incoming photon to be taken to a virtual energy state but the molecule here is already in a vibrationally excited state and hence the emitted photon has greater energy than the incoming photon and therefore appears at a shorter wavelength.

A typical Raman spectrum is shown in Figure 10.2. By convention, Raman spectra are shown with the intensity on the vertical axis as would be expected, but the horizontal axis is not normally labelled in wavelength but in inverse centimeters (cm⁻¹), in effect the energy change of the detected photon. This axis is therefore showing the spectral separation from the excitation wavelength and thus such spectra are independent of the excitation wavelength. A further feature is to the right hand side (i.e. further away from zero Raman shift) the greater the energy changes that have taken place in the Raman process. As the axis is independent of the excitation wavelength it makes spectra easier to compare and also makes calculations on the bond vibrational energies that have caused the spectral features. It is straightforward to convert a value in cm⁻¹ to nanometers as 1 nm is equivalent to 10,000,000/cm⁻¹. Thus 20,000 cm⁻¹ is equivalent to 500 nm.

The Raman scattering process is also affected, partially, by the polarization of the light. This is true even when the molecules are free to rotate randomly such as in a liquid, and the Raman effect is most clearly seen using an excitation source that is plane polarized. Thus, in scanning laser systems where the light is fibre delivered to the scan head, improved signal to noise on the image can be achieved using a polarizing component in the scan head.

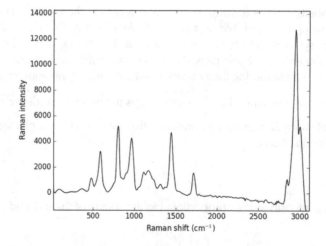

FIGURE 10.2 Typical Raman spectrum (PMMA spectrum, courtesy of Dr Penny Lawton, Durham University).

10.2 COHERENT ANTI-STOKES RAMAN SCATTERING (CARS)

As mentioned above, the Raman effect is generally weak and thus signals can be low, leading to long scanning times in beam scanned confocal microscopes or high power excitation which have an increased risk of damaging the sample. However, a non-linear-based Raman signal can be generated which is more selective and can thus provide greater signal to noise and higher speed imaging.

CARS is the third, and most recently developed, method of non-linear microscopy, following in the wake of multiphoton fluorescence microscopy (Chapter 8) and harmonic microscopy (Chapter 9). CARS is a four-wave mixing process in which a pump beam at frequency ω_p (with a wavelength λ_p) and a Stokes beam at frequency ω_s interact with a sample to generate an anti-Stokes signal at a frequency given by

$$\omega_{as} = 2\omega_p - \omega_s.$$

The CARS process was originally discovered in the Ford Motor Company but the acronym does not relate to its place of invention (Maker and Terhune 1965). As with the conventional Raman process described above, CARS can either be considered as a classical wave process, in a similar manner to that which harmonic imaging was explained in Chapter 9, or as a quantum mechanical photon process.

It is probably easier to gain an insight and understanding of the process using the quantum mechanical approach, which is illustrated in Figure 10.3. Here a photon of wavelength λ_p (frequency ω_p) is absorbed (where the higher energy level may be virtual) and then a second photon at wavelength λ_s drives the excited electron down to a real vibrational state. A third photon can now be absorbed sending the electron to another virtual state from which it is again instantaneously driven down, in this case to the ground state emitting light at the energy difference, which is then detected (λ_{detect}).

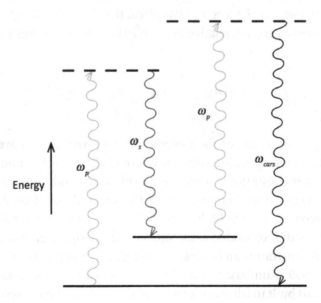

FIGURE 10.3 Energy levels present in a quantum mechanical approach to CARS.

If we consider the classical model, in line with equation 1 above, we need to consider the effect that the light's electric field has on the molecule's own electric field or the polarization of the molecule. In equation 10.1 we assumed that this effect was linear and only consists of one term. In fact the total polarization is an infinite sequence of terms in increasing orders of the electric field. Thus in line with equation 4 in Chapter 9 the expansion including the non-linear terms is $P_{total} = P_1^{(0)} + \chi^{(1)}E + \chi^{(2)}E^2 + \chi^{(3)}E^3 + \ldots$ For the Raman effect we consider changes in $\chi^{(1)}$ and in Chapter 9 we discussed $\chi^{(2)}$ and the effect with the square of the electric field in relation to harmonic imaging. In the case of CARS we consider $\chi^{(3)}$ and the electric field raised to the power of three (cubed). However, we have three laser beams in the CARS process and therefore the electric field is not a single cubic term. The three beams interact with each other through the $\chi^{(3)}$ term with the high peak powers of the electric fields driving the polarization of the molecules. By considering the molecular vibrations as a classical oscillator, with a frequency ω_v this interaction effect can be enhanced by driving the system at a resonance (i.e. the exact difference in the energy levels within the molecule). In CARS this oscillator is not driven by a single optical wave, but by the beat frequency $(\omega_p - \omega_s)$ between the pump and the Stokes beams. When this beat frequency is close to ω_v (where ΔE is the energy difference corresponding to ω_v in Figure 7) the molecular oscillator is driven very efficiently. It is the movement of the electrons within the molecule that give rise to the $\chi^{(3)}$ polarization term, and in effect this periodic motion is probed by a third beam (actually ω_p) whose frequency is altered by the addition of the beat frequency to produce the ω_{cars} (or λ_{cars} in terms of wavelength) emission at the so called anti-Stokes wavelength. Greater details on this process can be found in Boyd (2008).

In both approaches to understanding the CARS process the emitted light and driving of the various energy levels is undertaken coherently and thus all the individual molecules within the volume emit light at the same time and in phase, giving an enhancement in the signal, providing the *phase matching condition* is achieved. In order for constructive, rather than destructive interference to take place, the three wavelengths involved in the process need to have the correct relative phase relationship (the waves need to be in step). This is given by

$$\frac{n_{(\lambda_{detect})}}{\lambda_{detect}} = \frac{n_{(\lambda_p)}}{\lambda_p} + \frac{n_{(\lambda_s)}}{\lambda_s} \tag{10.4}$$

where n is the refractive index of the material (at the given suffix wavelength) and λ is the wavelength. In many optical techniques where CARS, or other non-linear processes are used, a significant constraint is placed on the physical conditions including the beam alignment (in frequency doubling for example the size and cut of crystals). However, in non-linear microscopy, due to the high numerical aperture of the lenses involved, the excitation volume has a radius of around a micron, not much longer than the wavelength of the light, and thus this constraint can be largely ignored. The first practical CARS microscope was described in 1999 (Zumbusch et al. 1999) and since then there have been multiple development systems built in laboratories around the world and more recently commercial microscopes.

As described above the process is driven by the $\chi^{(3)}$ term and the generated light is strongly emitted in the forward direction. This has implications on the exact optical configuration used, as will be discussed in Section 10.5. It is also important to note that the CARS process is significantly different from both fluorescent and even Raman microscopy in that there is no energy lost in the process and thus no heating of the sample. The downside of this is that in fact no resonant molecule needs to be present at all in the focal volume, leading to a non-resonant background. The removal of this background is a considerable practical complication, and the details can be found in Section 10.6. There is, however, no risk of the molecule becoming damaged in the process in the manner of photobleaching in fluorescence.

10.3 STIMULATED RAMAN SCATTERING (SRS) MICROSCOPY

Although CARS is a very powerful non-linear imaging method there are certain drawbacks. The first consideration is the non-resonant background term which affects the signal to noise ratio; the second is the spectra that are determined (by the tuning of the lasers) are not always those expected in conventional Raman spectroscopy and the results can be confusing. A variation of CARS, in that it is a non-linear process using Raman transitions, is stimulated Raman scattering (SRS) (Freudiger et al. 2008). In this technique the two laser wavelengths are tuned such that their beat frequency is close to a Raman vibration. When this beat frequency exactly matches the Raman frequency of a molecular vibrational mode, amplification of the Raman signal is achieved by what can be viewed as "stimulated emission". The pump beam can be thought of as exciting a mode up to a virtual energy state and it is then driven back down to a state close to its original energy state by the "Stokes" laser beam. This state will have a slightly higher energy than the original ground state with the energy difference being the vibrational mode or Raman frequency.

As the virtual state is driven down by the Stokes beam one now has in effect two Stokes photons (the stimulating photon and the energy from the excited state decaying down), the loss of a pump photon and a molecule now in a slightly higher vibration level. The imaging signal can thus be detected either by looking for a loss in intensity of the pump beam (stimulated Raman loss, SRL) or an increase in intensity of the Stokes beam (stimulated Raman gain, SRG). If the frequency difference between the two incoming beams does not match a vibrational mode, the excited molecule is not driven down to the lower energy state, and therefore there is a non-background signal. This process relies on the intensity of two light sources and is thus again non-linear. Although the non-resonant background seen in CARS is not present, one is now looking to detect a small change in intensity on a signal that comes directly from the laser, leading to practical challenges.

10.4 PRACTICAL RAMAN MICROSCOPY INSTRUMENTATION

At the core of a Raman microscopy system is the requirement to deliver a laser to a specific spot on the sample and to then collect the Raman signal for subsequent detection. However, as can be appreciated from the description of the physical processes that take place to generate the Raman signal, the details, and practical implementation, of these core requirements vary depending on the exact Raman process being used to generate the contrast in the image. In general though, practical Raman microscopes use a beam scanning technique to generate images and the core configuration is similar to that used in a confocal microscope.

The basic beam scanned Raman microscope is illustrated in Figure 10.4. Most modern Raman systems contain a laser source operating at around 785 nm, or 532 nm. There are two conflicting requirements on the selection of the excitation wavelength. The first is that the size of the Raman signal depends roughly on the wavelength to the fourth power (λ^4) and thus the shorter the wavelength the greater the signal. However, fluorescence is a major complication to the sensitive detection of Raman signals, as the fluorescence process is much more efficient than Raman generation, and in general, samples have the potential to be fluorescent when excited at shorter wavelengths. Thus a frequently used wavelength in modern Raman microscopes is at around 785 nm. Here a laser diode can be used, helping to keep the cost of instruments lower, and also helping to maximize the Raman to fluorescent signal ratio. The other common wavelength is at 532 nm where a frequency doubled diode pumped Nd:Yag laser provides the excitation. In biological samples this does, however, increase the probability of background fluorescence signals though methods to

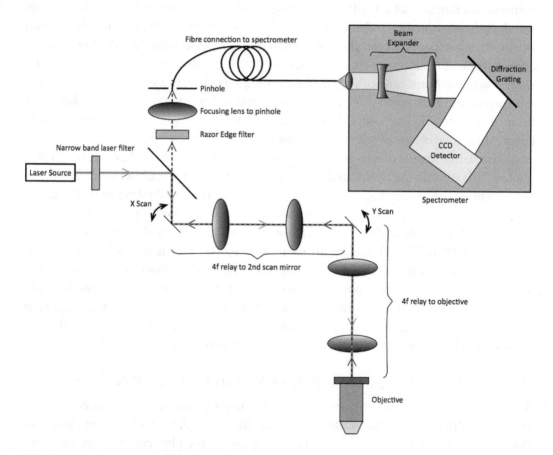

FIGURE 10.4   Beam scanned Raman microscopy.

mitigate against these are described later. Other laser wavelengths will clearly work for Raman microscopy but are less commercially viable.

Beyond its wavelength the actual requirements on the laser are slightly more stringent than those for conventional fluorescence microscopy. In the case of fluorescence the exact wavelength of the excitation laser, in general, does not matter as the absorption, and crucially emission, wavelengths of fluorophores are very broad. Thus if the wavelength of the excitation source drifts slightly (even several nanometers) this will not create a significant problem. The situation is very different in the case of a Raman microscope where the emitted light has to be examined using a spectrometer and the exact wavelengths of the detected signal is used to determine what is present within the sample. A Raman shift of 1000 cm^{-1} corresponds to a wavelength change of 66 nm for excitation at 785 nm. However, a change in wavelength of 1 nm for the excitation source corresponds to a Raman shift of 10 cm^{-1}. Frequently, when looking for specific chemical features within a sample, one is interested in such small changes and thus the excitation laser needs to have a wavelength stability of better than around 0.1 nm. Laser diodes are now suitable for use as such excitation sources though care has to be taken to ensure sufficient wavelength stability. Typically the excitation laser will have a power of up to 100 mW though the actual power on the sample is significantly lower than this, albeit higher than the typical power used for fluorescence-based imaging. As with fluorescence microscopy the laser intensity must be stable with a noise of significantly less than 1% of the full intensity, and ideally around 0.01%.

In the typical system shown in Figure 10.4 the laser frequently passes through a narrow band filter to ensure that no broad spectral light from the laser (amplified spontaneous emission) reaches the sample and hence detection system. Although orders of magnitude lower in intensity than the laser line, this pedestal of light can extend over a broad spectrum and provides a background on the signal. The narrow spectrum of light then passes through a specialized dichromatic beam splitter before entering the beam scanning system, to be delivered to the sample in a manner identical to that in a confocal microscope. The specialized filter is known as a *razor edge* filter which only transmits a narrow band (typically 2–3 nm) centred on the laser wavelength. This ensures the maximum signal is sent to the detector on the return path. The beam scanned light then reaches the objective lens on the microscope. In most Raman systems the objective is an air objective, minimizing the risk of unwanted background signal from any immersion fluids that would be required with oil or water objectives.

The returned light then passes back through the scanning system and is reflected off the dichromatic mirror and through a further narrow band laser blocking filter before reaching the detection system. In order to obtain Raman spectra for each point on the sample the light from each pixel could be sent through a spectrometer in which a grating is scanned and the spectrum recorded; however, all systems now use a fixed spectrometer grating the output spectrum of which is sent onto a CCD camera array. The exact type of detector used depends on the spectral resolution required and also the light source being used for the excitation. As the sensitivity of silicon falls off rapidly at wavelengths longer than around 950 nm, thus using a 785 nm source, the maximum practical Raman shift that could be detected is around 2700 cm⁻¹. The selection of the exact grating and camera combination to be used is discussed in Section 10.6.

It is also clear that the Raman spectra can be recorded with a confocal pinhole included in the detection system before the spectrometer. Here the signal light reflected from the dichromatic beam splitter is focused down through a pinhole before being re-collimated and passed into the spectrometer system. The advantage of a confocal system here, beyond the obvious ability to collect three-dimensional data sets, is that the pinhole can be used to reject Raman signals that might be coming from any immersion oil used or from the sample mount as even glass has a detectable Raman signal. The clear disadvantage is that one is taking the signal from a smaller volume and thus there is a loss of signal leading to extended imaging times to provide clear enough spectra.

The data sets recorded then have to be analyzed and presented. For a two-dimensional image this is typically achieved by storing the data as a 3D data set (Figure 10.5a). For each pixel recorded the spectrum is saved as the third dimension and then by selecting a pixel, or group of pixels, the spectrum for a particular region can be shown (Figure 10.5b). It is also possible to encode certain spectral features in a specific colour to bring out the spatial distribution of a particular molecule. Such data sets clearly can become very large, for example for a 512×512 pixel image in which each pixel has a spectrum in say 300 spectral "bins" (10 cm⁻¹ bands up to 3000 cm⁻¹) and each spectral point is recorded with 8-bit accuracy (0–255 intensity values) the image will be in excess of 75 Mb. If one then has a 3D data stack it can be appreciated that file sizes can grow quickly and this should be appreciated when spectra are being collected. This is discussed in Section 10.6.

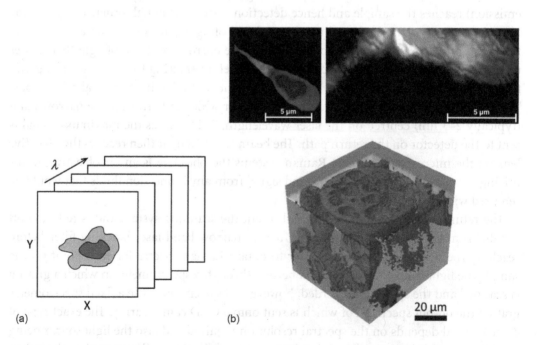

FIGURE 10.5 a) A diagrammatic representation of the data storage in Raman microscopy; b) 2D and 3D Raman images and data stacks (courtesy of Renishaw and Horiba).

10.5 PRACTICAL CARS MICROSCOPY INSTRUMENTATION

A typical CARS system is illustrated in Figure 10.6. In order to achieve CARS excitation one needs two laser sources tuned to the correct wavelength with sufficiently high power to achieve a detectable signal. Although in principle it is possible to achieve this using a continuous laser, all practical CARS microscopy has been undertaken with short-pulsed lasers. In the system illustrated the output from a mode-locked laser has been sent into an optical parametric oscillator (OPO). This OPO uses non-linear optical techniques to produce a tunable and intense light source. The second laser pulse is typically the initial laser used to seed the OPO. The OPO output is then tuned to ensure that the wavelength combination drives the required transition in the molecule of interest.

As with multiphoton and harmonic microscopy one requires a high peak power from the light source to maximize the signal, but a low average power to minimize potential damage to the sample. In the case of multiphoton and harmonic imaging the best solution is to use a laser with a pulse length of around 100 femtoseconds, as described in Chapter 8. However, the requirements for CARS (and other non-linear microscopy methods) are different. A 100 fs pulse at 800 nm has a spectral bandwidth of around 16 nm, which means that it will excite multiple Raman modes rather than the one specifically targeted. Using picosecond laser pulses, however, the spectral bandwidth falls significantly (to less than 1 nm) and so one has excellent spectral overlap to individual Raman features (Ganikhanov et al. 2006). The remaining challenge with using two ultra-short pulse lasers is to ensure

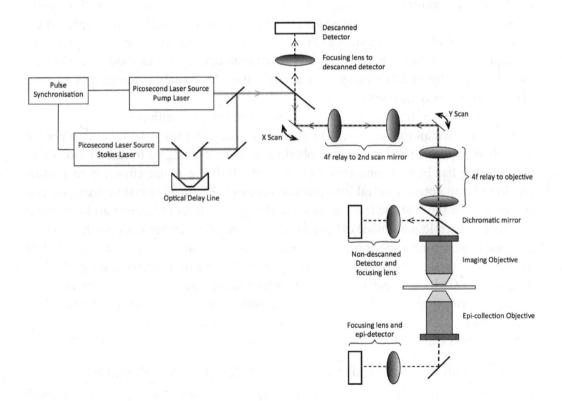

FIGURE 10.6   Beam scanned non-linear Raman microscopy.

temporal overlap of the light within the sample volume. For two picosecond pulses this means path length differences between the pulses must be reduced to less than 300 μm. Using a single laser to initially generate the pulses also helps with regard to temporal overlap, as pulse-to-pulse jitter in timing between the pulses does not affect the signal.

As with the other non-linear microscopy techniques, as the excitation volume is inherently small within the sample, one does not require a pinhole in the optical system to generate three-dimensional images. Thus the returned light, which will be at a shorter wavelength than any of the excitation beams, can be collected without having to pass back through the scanning optics. In fact the majority of the CARS signal as noted above is generated in the forward direction and thus using a second objective mounted after the sample can provide the largest signal in a configuration known as *Forward CARS* (F-CARS). A spectrometer is not needed in a typical CARS system as the spectral selection of the Raman mode of interest is achieved through the tuning of the lasers.

Despite the excitation being localized in space there is, however, a non-resonant background signal in CARS imaging. Technically this is due to the electron, rather than vibrational, contributions to the $\chi^{(3)}$ parameter. The magnitude of this non-resonant background is dependent on the wavelengths of the excitation lasers used with visible wavelengths leading to a larger contribution. This is a further reason for using near infrared lasers to excite the CARS signal. In addition, as in the case of multiphoton microscopy, the longer wavelengths also suffer less scattering within the sample enabling greater imaging depths. The non-resonant background can limit the sensitivity of the CARS methods. Although the CARS signal propagates mainly in the forward direction, in samples with a reasonable level of scattering, such as brain tissue, the epi-configuration is preferred with the signal being scattered back through the excitation optics. This method is particularly suitable when objects being imaged are smaller than the excitation wavelength as more light is directed back through the system.

As with all beam scanned methods, the speed of imaging within CARS, and indeed conventional Raman microscopy, is limited by the scanning speed and the time needed to dwell at a specific pixel in order to obtain a good enough signal to noise ratio. Video rate imaging has been demonstrated (Saar et al. 2010) but is not currently in routine use. In order to obtain spectral information from a CARS imaging system there are two approaches. In the first the laser is tuned to the specific lines of interest and the image recorded for each Raman mode of interest. This means one then has multiple images with each one representing a specific Raman feature. The alternative is to dwell at a pixel while the laser is scanned in wavelength recording the CARS Raman spectrum for each pixel. In general in CARS microscopy one knows which Raman bands are of interest and thus tuning the laser to specific wavelengths is the method of choice so that (a) the sample is only exposed to laser light when a signal is expected, and (b) the imaging is speeded up as images are only captured when a signal is expected.

10.6 TECHNIQUES AND APPLICATIONS IN RAMAN MICROSCOPY

We now need to turn to the practical details that have to be considered in using Raman microscopy. Before any imaging is undertaken the wavelength calibration of the system should be undertaken. In commercial systems this is normally a standard procedure run using a start-up routine. If this is not present a highly Raman active sample with a known spectrum can be placed under the microscope to confirm the calibration. Toluene, for example, has strong Raman peaks at 786, 1003 and 1030 cm^{-1}. Other solid samples used include polystyrene beads.

The main benefit of Raman microscopy is that the contrast is provided by the chemical bonds present within the sample, and thus one does not need to introduce an exogenous compound as is generally the case with fluorescence microscopy. This means that there is less chemical perturbation to the sample, providing there are compounds present that are Raman active. The downside of this is that the method can be less specific and in the case of Raman microscopy in biological samples the images can be hard to interpret due to the complex spectra produced. With advances in computer analysis and crucially data reduction tools, the complexity of the spectra has become less of an issue. However, the fact that many materials are Raman active does determine some of the major considerations for sample imaging.

The first consideration is in the way that the sample is mounted. Generally most plastic materials that might be considered as sample holders are Raman active, indeed polystyrene beads are often used as a test sample. This leads the users towards glass slides, used in most microscopy applications; however, glass is also Raman active giving a broad spectral background (Figure 10.7b). The use of quartz slides reduces this background signal and crystalline quartz is a further improvement, however this adds significantly to the cost of the consumables for Raman imaging. The situation is better in the case of confocal Raman microscopy as the glass signal can be rejected using the depth sectioning capability of such a microscope. The positive advantage of using a glass slide is that one can use the glass signal to ensure that the system is working correctly. For fixed biological samples one should also consider both the fixing and sectioning process used, minimizing the use of Raman active materials that could still be present when the sample is imaged.

With the sample prepared one then needs to consider the use of air objectives ahead of the more commonly used, and higher NA, oil objectives. The higher NA lens will have (a) higher lateral resolution, (b) higher signal gathering power, and (c) greater depth sectioning capability, but any immersion fluids used will tend to produce a background signal. The first lens choice is thus normally the highest NA air objective available. The lens should also be checked that it is suitable for transmitting light at the expected emission wavelength, which will frequently be around 900 to 1000 nm for 785 nm excitation. In addition the use of many objective lenses at these longer wavelengths can lead to a loss of signal due to chromatic aberration. As discussed in Chapter 3, objective lenses are optimized at a range of wavelengths but by convention these are in the visible portion of the spectrum. It is thus worth consulting with the microscope manufacturers which lenses from their range they suggest are the most suitable for Raman microscopy.

With the sample therefore mounted in the best material, and the most suitable lens selected, one then needs to consider the trade-off between the spectral resolution required, the signal to noise in the image and the imaging time. For high spectral resolution the detected light is spread using a diffraction grating across the CCD detector. The greater the spectral spreading (i.e. the higher the spectral resolution) the fewer photons that will arrive

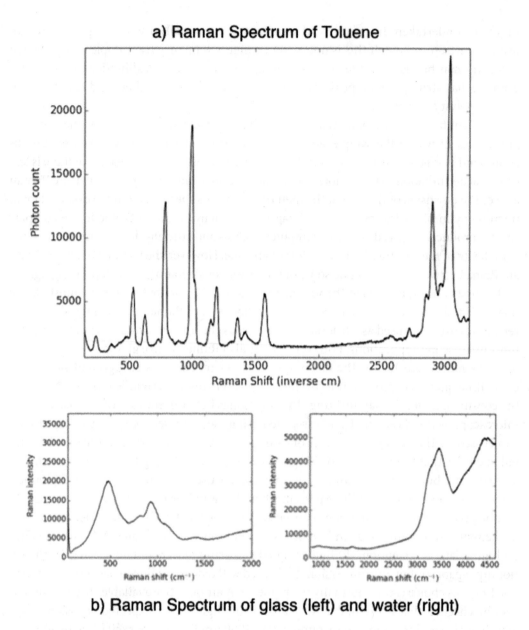

FIGURE 10.7 a) Raman spectrum of toluene, b) Raman spectrum of glass and water (Spectra courtesy of Dr Penny Lawton, Durham University).

on each pixel; hence one either has to live with a poor signal to noise or use a long exposure on the camera for each pixel. An alternative is to record the images with a lower spectral resolution by "binning" pixels on the camera. This enables their signals to be added, reducing some aspects of the noise. It is common to record the first images at a lower spectral resolution before increasing the resolution and using a longer integration time for the images that will be analyzed in detail. It should also be remembered at this point exactly why Raman is being used for a particular imaging task. If the main aim is purely as a contrast mechanism to obtain structural images of the sample then relatively crude spectral resolution may be sufficient to separate out the physical features of interest. If, however, one is after specific chemical and spatial information then the higher resolution, longer imaging time approach is likely to be required.

Typically before capturing the data of interest an image is taken in an area of the sample where there are no features. This background image contains the spectral features from say the glass slide. This "blank" image can then be subtracted from the real images to remove unwanted background Raman signals. This operation should be undertaken with care and it is always better practice to record an image and undertake the background subtraction as a post imaging procedure so that the raw images are always available.

Although there is no Raman equivalent of fluorophore bleaching the laser power used in Raman microscopy can damage the sample. In line with all optical imaging the illumination power level should therefore be kept as low as possible to provide sufficiently good signal to noise images without sample damage. In taking 3D confocal data stakes it should also be remembered that slightly higher excitation power may be required to obtain a good signal to noise ratio as the collection volume for the signal will be small.

10.7 TECHNIQUES AND APPLICATIONS IN CARS AND SRS MICROSCOPY

Due to the complexity of a non-linear Raman microscope the first task should always be to place a known, highly Raman active sample on the microscope stage to optimize the laser alignment to ensure perfect spatial and temporal overlap of the beams in the focal volume. The comments made above in relation to confocal Raman imaging on sample mounting and preparation are equally applicable to CARS and SRS imaging. In addition to these Raman specific comments the practical details in Chapter 8 on multiphoton imaging are also useful when undertaking CARS microscopy. There are, however, a few non-linear Raman techniques that can be used to improve the image quality and usefulness of CARS microscopy. These mainly focus on methods of improving the signal to noise using different light collection and detection methods.

As discussed above, the CARS process inherently sends the light in the forward direction and thus in many samples consideration should be given to collecting the light travelling in the forward direction (away from the illuminating objective). As has been discussed in Chapter 8, there are several approaches that can be used. The most obvious one is to use a detector that collects the light travelling in the forward direction. To do this most efficiently a collection lens is required but this is already present in many microscopes as one can use the illumination condenser lens. If a dichromatic beam splitter is then mounted behind this lens the light can be sent straight to a detector. Alternatively a second objective lens can be used to collect the emission before again being directed to a suitable detector. In both cases descanning of the beam is not required as the excitation is localized due to the non-linear processes involved in generating the signal. Aside from the optical collection system the other challenge here is to place blocking filters in the system to remove all the residual laser excitation light. As one may be using several 10s of milliwatts of excitation power and the signal may perhaps be in the nanowatt level, the blocking needs to be very high and frequently several narrow band filters are used.

A potentially simple alternative to using a transmission detector is to place a mirror above the sample to direct the light back towards the illumination objective. Ideally this mirror would be placed after a collecting lens so that the emission is sent back to the optimal point for collection by the primary objective. However, even a simple plane mirror (or even reflectively coated sample slide) helps in redirecting the light. In more highly scattering samples, such as the brain discussed above, the emission light will be highly scattered and deep imaging can be achieved using a detector mounted before the scan head to collect light backscattered towards the excitation objective. As was discussed in Chapter 8 this detector should be placed as close to the back aperture of the objective as possible, minimizing the number of optical elements that the light passes through and also ensuring the light, which will emerge at a range of angles as it will have been scattered, has a maximum chance of reaching the detector.

The other improvement that can be made, in particular for deep imaging, is to consider the use of adaptive optics to improve the excitation volume when this is deep within the sample. In previous work, signals as deep as 750 μm have been increased by up to ten times using adaptive optic beam shaping on the input to the microscope (Wright et al. 2007). The advantage, beyond just being able to image more deeply, is that lower excitation powers can be used. In CARS microscopy, when one laser is frequently operating beyond 1000 nm, there is a risk of localized heating due to laser power being absorbed in the water present in most biological samples. The method also ensures that close to diffraction limited images can be obtained from deep within the sample.

The other non-linear Raman method described above is that of stimulated Raman scattering. Here the optical configuration is typically slightly different from the conventional CARS system (Freudiger et al. 2008). Figure 10.8 shows the equipment configuration now normally used in SRS. As discussed above, in SRS one is looking for a slight change in intensity on one of the excitation laser beams, either an increase (stimulated Raman gain, SRG) in the Stokes beam or a loss of intensity (stimulated Raman loss, SRL) in the pump beam. The best method to detect a slight change in an otherwise high intensity signal is to modulate the signal (by changing the intensity of the other beam) and then using a lock-in,

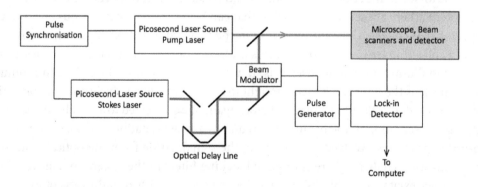

FIGURE 10.8 Optical configuration of lock-in detection for stimulated Raman scattering microscopy.

or phase sensitive, detector designed to compare the signal with and without the chopped beam. Even then one requires a detector with a high dynamic range as one is looking for a small signal change.

10.8 WHEN TO CONSIDER THE USE OF RAMAN MICROSCOPY

As was stated at the start of the chapter Raman-based microscopy has shown a significant growth in the last few years though the instrumentation has not been made by the major microscopy companies, in general, but by spectrometer companies that add a microscope to the front of their spectrometers. The main advantages of Raman imaging are its label-free nature and if the sample has sufficient concentrations of Raman active components then this provides an excellent route to imaging a wide range of samples. The main drawback, compared to say fluorescent labelling microscopy, is that one may image a specific compound, by selecting its spectral feature, but one then knows where that compounds is and perhaps not some of the rest of the structure of the sample. In the case of biological samples this can mean that one is monitoring the protein output from a cellular operation, but one cannot be certain that the chemical is still in the area in which it was produced. The signal levels are also in general low compared to fluorescence-based imaging and the equipment more expensive. There are, however, a growing number of Raman "libraries" linking spectral features to specific activity within the cell, and with the constant increase in computing power, matching a specific spectrum to a process within a cell is becoming increasingly possible. One particular area of interest here is in potential routes to cancer detection, perhaps removing the time-consuming requirement for H&E staining in the pathology laboratory.

CARS or SRS adds the ability to develop three-dimensional images and to make the image very chemical specific. Here some labelling has been undertaken with biological samples being loaded with deuterated compounds, which have a slightly different Raman spectrum (due to the different mass of deuterium in the vibration modes). This means that dynamic changes in cells can be monitored as a specific C-H vibration changes to C-D as the sample undergoes normal biological processes (Xie, Ji and Yang 2006). The method has also been applied to tracking non-fluorescent pharmaceutical drugs through tissue samples, in particular to assess transdermal drug delivery.

Raman methods are probably, at present, underutilized but the growing interest and data available to interpret spectra make these microscopic techniques that should be considered if one wishes to minimize perturbation to the sample.

REFERENCES

Boyd, R. W. 2008, *Nonlinear Optics*. 3rd edn. Elsevier.

Freudiger, C. W., W. Min, B. G. Saar, S. Lu, G. R. Holtom, C. He, J. C. Tsai, et al. 2008. "Label-Free Biomedical Imaging with a High Sensitivity by Stimulated Raman Scattering Microscopy". *Science* 322, 1857–61.

Ganikhanov, F., S. Carrasco, X. S. Xie, M. Katz, W. Seitz and D. Kopf. 2006. "Broadly Tunable Dual-Wavelength Light Source for Coherent Anti-Stokes Raman Scattering Microscopy". *Optics Letters* 31(9): 1292–4.

Maker, P. D. and R. W. Terhune. 1965. "Study of Optical Effects Due to an Induced Polarization Third Order in the Electric Field Strength". *Physical Review* 137(3A): A801–18.

Saar, B. G., C. W. Freudiger, C. M. Stanley, G. R. Holtom and X. S. Xie. 2010. "Video-Rate Molecular Imaging in Vivo". *Science* 330: 1368–71.

Wright, A. J., S. P. Poland, J. M. Girkin, C. W. Freudiger, C. L. Evans and X. S. Xie. 2007. "Adaptive Optics for Enhanced Signal in CARS Microscopy". *Optics Express* 15(26): 18209–19.

Xie, X. S., Yu J. and Wei Y. Y.. 2006. "Living Cells as Test Tubes". *Science (New York)* 312(5771): 228–30.

Zumbusch, A., G. R. Holtom and X. S. Xie. 1999. "Three-Dimensional Vibrational Imaging by Coherent Anti-Stokes Raman Scattering". *Physical Review Letters* 82(20): 4142–5.

CHAPTER 11

Digital Holographic Microscopy

D IGITAL HOLOGRAPHIC MICROSCOPY (DHM) is a relatively new area of optical microscopy that is now making significant advances due to improvements in both electronic cameras and more specifically low-cost high-performance personal computers. The
method uses small changes in the optical path of light passing through, or being backscattered from, the sample as the contrast mechanism in the image. There is therefore some
similarity with DIC, phase contrast and interference methods as there are no requirements
to add chemicals or perturb the sample in order to obtain high-contrast images. The other
key feature is the ability to record a full three-dimensional image in a single shot. It is this
difference that really marks the method out from DIC imaging. Images from different
optical depths can be recovered from the recorded hologram through the use of advanced
computation techniques. It is actually possible to record images without the use of the
lens, and compared to other optical microscopy methods the resolution is less determined
by the perfect optical quality of the components and more by the number and size of pixels on the recording device. At present the method is gaining greater acceptance within
the field of materials, micro-components and surface science than in the life sciences.
This is probably because of the desire in the life sciences to image using fluorescence, or
Raman-based chemical signals, which can more easily be attributed to a particular cellular
process or feature.

The term holography was first used in 1948 by Dennis Gabor, the inventor of the technique (Gabor 1948, 1949) when he was looking at methods to improve electron microscopy,
in particular the aberrations caused by the electron lenses used in the system. The word
holography comes from two Greek words, *holos* (whole) and *graphe* (drawing or writing).
The method records all the information about the incoming light field with both its amplitude and phase, and is based upon interference. A patent (1947) protected the original
work but as with many such breakthrough ideas the technology was not available to exploit
the advance. Real practical application and development of holography had to wait for

the invention of the laser. With the laser's long coherence length Emmett Leith and Juris Upatnieks produced the first really practical optical holograms, in particular introducing an off-axis method of illumination (Leith and Upatnieks 1962, 1963). For his original idea on holography Gabor was awarded the 1971 Nobel Prize in Physics.

The chapter initially discusses the basic physical principles behind the method and this is followed by the most recent practical implementations covering both the optical and computational aspects of the holographic method. In particular, with regard to the computation techniques, there are a number of references given as this area of the field is changing very rapidly. As computer power increases and even low-cost computers can carry out complex calculations on multi-megapixel images very rapidly, and new software developments are taking place very rapidly, the reader is advised to search for the most recent publications. The chapter ends with a consideration of the areas in which digital holographic microscopy is currently most widely used and those areas in which the potential for future applications is greatest.

11.1 PHYSICAL BASIS OF THE METHOD

At the core of the holographic method is the recording of both the intensity and phase of the light that passes through the sample. In order to achieve this the illumination light is split into two as shown diagrammatically in Figure 11.1. One beam passes through the sample and the second, reference beam, travels "around" the sample and is subsequently recombined with the sample beam on the digital camera sensor where the resulting interference pattern is recorded. The recorded pattern of light is then processed by a computer. Here the exact optical differences in the light paths are calculated and hence the sample's optical properties can be reconstructed as an image. There are many variations in which slightly different optical and computational methods are used, but all systems at their core have a coherent light beam being divided, taking two optical paths and then being recombined.

In Figure 11.1 the light from a laser source is initially passed through a spatial filter, to ensure a high-quality single spatial mode light profile. This improves the overall image quality by reducing intensity variations in the beam as these can lead to noise in the final image. Although lasers are most commonly used, due to their high temporal coherence, LED illumination is possible though the optical path length difference between the light

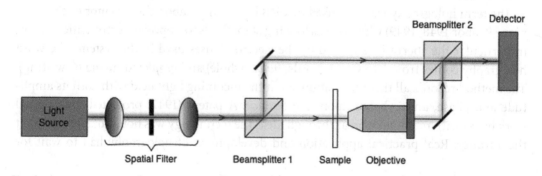

FIGURE 11.1 Basic optical system for digital holographic microscopy.

passing through the sample and the reference beam then need to be matched to around 10 µm. The "clean" light beam is then split into two parts with one passing around the optics of the microscope to provide the unaltered reference beam. The remaining beam is then directed onto the sample as a collimated light beam. In transmission, the most common configuration, the light passing through the sample is then normally collected by a high numerical aperture objective lens. Though the NA of the lens does play a role in the final image resolution the actual resolution is more strongly determined by the number and size of the pixels on the detector and the algorithm used in the image reconstruction. Indeed it is possible to record a high-resolution image even without a lens collecting the light after the sample (Isikman et al. 2011). The two beams are then recombined using a second beam splitter before reaching the detector where the interference pattern is recorded.

Figure 11.2 illustrates what happens to the two light beams as they each make their way to the detector. The light of the reference beam passes through to the detector unperturbed. For the light passing through the sample three things can happen:

1. The light can be scattered, or absorbed and hence not pass through the rest of the optical system (Figure 11.2a).

2. The light can pass through the microscope slide and surrounding fluid (Figure 11.2b).

3. The light can pass through the microscope slide, fluid and also the sample (Figure 11.2c, d).

As the light makes its way through the imaging arm the different parts of the wave front have different optical path lengths. As discussed in Chapter 2, as light passes through a material it is slowed down (relative to a vacuum) with the amount of slowing being determined by the refractive index. This effect is demonstrated in Figure 11.2c, where the light is delayed by different amounts due to the physical path length, and in 11.2d, where the light is delayed by different amounts due to differences in refractive indices. These differences in path length then appear on the detector, when recombined with the reference beam, as intensity variations in the image. The intensity of the image is determined by the amount of delay (its phase relative to the reference) and also the number of photons lost (its amplitude relative to the reference). The interference pattern recorded thus has all the information required to reconstruct the path of the light through the sample, and the computer's role is to determine this path for each part of the light beam. It is also possible to configure the microscope to operate in a backscattered mode in which the light is passed back towards the source, where the optical considerations of the light beam are similar to those discussed above.

The mathematical treatment of the light travelling through the sample is beyond the scope of this book, even as a technical inset. For those interested there is an excellent open access review of the different methods by Myung Kim (Kim 2010) and in a book by the same author (Kim 2011). The key points are that the passage of the light can be completely reconstructed using a range of algorithms, and that a computer generated lens placed into the reconstruction, which can be focused to different depths within the sample, enables

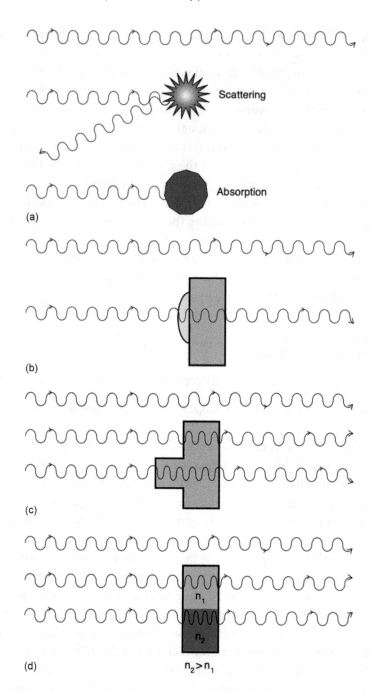

FIGURE 11.2   Light paths through the sample region of a digital holographic microscope compared to the reference beam: a) scattering and absorption, b) through the microscope slide and surrounding fluid, c) through two different thicknesses of sample, d) through a region of the sample with different refractive indices.

optical sections to be taken. After reconstruction one has a conventional bright field image as well as an image which demonstrates the phase shift received by each part of the light beam.

11.2 PRACTICAL IMPLEMENTATION

Figure 11.3 shows the two basic configurations for a digital holographic microscope. 11.3a is a configuration suitable for a reflective, or predominantly backscattering, sample most commonly found in materials research, whilst 11.3b is that used for a predominantly transmissive sample, typically biological specimens. The first configuration is known as

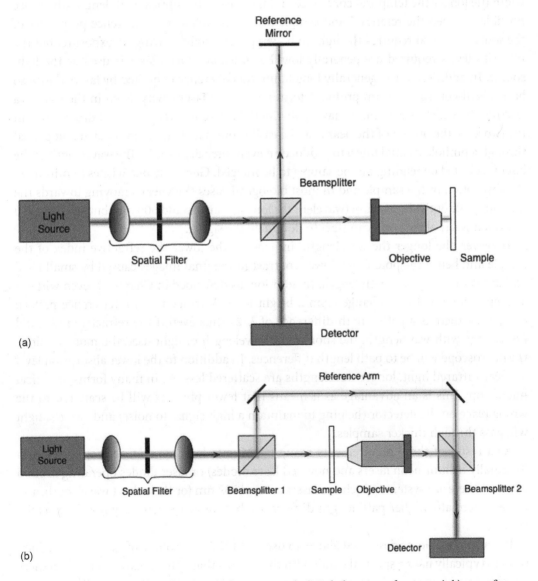

FIGURE 11.3   a) A configuration for a reflective sample (Michelson interferometer); b) a configuration for a transmissive sample (Mach-Zender interferometer).

a Michelson interferometer and that for the transparent sample a Mach-Zender interferometer. Once the type of sample to be investigated is known the main considerations in selecting, or designing, such a microscope centre on the choice of light source, the detector and the subsequent processing algorithm.

11.2.1 The Light Source for Holographic Microscopy

As has been discussed above, the light source has to have sufficient temporal coherence to enable the reference beam and sample beam to be recombined and to still be coherent for the interference to take place. The temporal coherence, as discussed in Chapter 2, is directly related to the spectral bandwidth of the source. The narrower the spectral bandwidth the longer the temporal coherence and the greater the optical path length difference possible between the reference and sample beam. As well as the coherence properties of the source one also requires the light to be temporally stable during an exposure, but the intensity levels required are generally low. For most systems a laser is used as the light source. In addition to the generally long temporal coherence exhibited by lasers they also have a well-controlled beam profile determined by the laser cavity. Even in the case of a laser diode, which is astigmatic having distinctly different divergence and dimensions in the two axis, the output of the beam can be easily collimated and manipulated and passed through a pinhole (spatial filter) to produce an even intensity profile. The wavelength of the laser is selected depending on the sample to be imaged. Generally one wishes to minimize the absorption in the sample and in most biological cases this means moving towards the red end of the optical spectrum (wavelengths longer than around 600 nm) but less than 950 nm when water absorption can start to dominate the signal.

However, the longer the wavelength, in general, the lower the refractive index of the sample and hence the potentially lower contrast in the final images caused by small local variations in the optical path length. In addition, as described in Chapter 2, even without a sample there will be a change from a bright to dark fringe in an interference pattern every time there is a path length difference of $\lambda/2$. Thus even if the refractive index did not change with wavelength, the shorter the wavelength of light used the more sensitive the microscope will be to path length differences. In addition to the lower absorption level for near infrared light, longer wavelengths are scattered less. As in many forms of optical microscopy this is an advantage as it means that fewer photons will be scattered to the wrong place on the detector (helping to maintain a high signal to noise) and also the light will pass through thicker samples.

As a result of these requirements many DHMs use lasers operating around 630 nm (originally helium-neon lasers and now red laser diodes) or laser diodes operating around 780 nm. There are systems which use lasers around 532 nm for their short wavelength, and hence potentially higher path length differences, but these are generally not commercial systems.

It is also possible, as described above, to use an LED as the source of illumination. Such sources typically have a spectral bandwidth that means that optical path length differences between the reference and sample beams are restricted to less than 10–20 μm. The collimation and beam shaping optics are also more complex. However, LEDs have an inherently lower cost than laser diodes and one can therefore design a system with multiple wavelengths that can be used to produce sets of images enhancing different features within a sample.

The other consideration in the source is control of the polarization of the light. In general, the laser sources discussed are polarized and with the insertion of an additional polarizing element in the reference beam the intensity of the reference and sample beam can be equalized. This helps to maximize the recorded signal to noise ratio (contrast in the hologram). Using polarization optics it is also possible to undertake birefringent imaging of the sample to provide an additional contrast mechanism. This is often very useful if one wishes to observe stress patterns within a sample or to enhance local gradients in the refractive index.

11.2.2 The Detector for Holographic Microscopy

Probably the most important consideration in a DHM is the detector. In most microscopy systems a major concern about the detector is the sensitivity in order to maximize the signal to noise. Although this is clearly relevant in DHM the pixel size and number of pixels are more important as they are the main factors in determining the ultimate resolution of the microscope. They have to record the hologram with as great a resolution as possible as the detail in the image is recorded in the high spatial frequencies in the interference pattern. If that detail is missing no level of subsequent computer processing can return the fine structure. The details on the types of detector systems used in DHMs are described in Chapter 2, but normally consist of CCD or CMOS cameras. Here, there is sometimes a trade-off between pixel size (sensitivity) and the resolution that can be captured from the interference pattern.

The full details on the effect of the pixel size and number, and the position of the detector, can be found in three publications to which the interested reader is referred for specific details (Wu and Gao 2015; Hao and Asundi 2011; Kelly 2009). However, an indication of the roles of the different factors can be seen through a series of approximations.

The lateral resolution of a system is given by

$$\Delta x_i = \frac{\lambda d}{M \Delta x} \text{ and } \Delta y_i = \frac{\lambda d}{N \Delta y}$$

where λ is the wavelength of the light, d is the distance between the object and the detector, M and N are the number of CCD pixels along the x and y directions respectively with Δx and Δy being the pixel sizes in those directions (Kelly 2009). Thus it can clearly be seen that to maximize the resolution (make Δx_i and Δy_i as small as possible) one needs a large number of small pixels, at a short wavelength and with the smallest distance possible from the object to the detector. This simple formula is further complicated when the effect of a collection lens or train of optics is included, but a simple practical example provides an indication of the actual resolution that can be achieved.

As such a practical example we can consider a holographic system similar to Figure 11.3b using a helium-neon laser operating at 632.8 nm imaging a 0.95 μm polymer bead. For a

CCD with a pixel size of 4.65×4.65 μm and 1392×1042 pixels in combination with a ×40, 0.55 NA objective and recording the hologram 140 mm from the sample, the beads have an apparent size of 1.337 μm, or about 1.4 times the actual size of the bead. If this system was used in a conventional mode then the diffraction limit for this lens is around 0.64 μm. When this is convolved with the bead size one would expect an imaged size of around 1.2 μm, so we are slightly away from the diffraction limit. However, in taking the hologram when the image is correctly processed we are able to record a full three-dimensional image of the beam, which would not be possible with a conventional microscope.

The final consideration is to use a monochromatic detector. This means that each pixel has the same level of sensitivity to the incoming light. In the case of most colour cameras this is the Bayer filter described in Chapter 2. This spectral filter in front of the pixels means that the individual pixel intensity is not just a function of the light path and resulting interference. There are mathematical methods by which the intensity can be determined either by applying a correction factor for each pixel or by combining pixels, which means a loss of resolution as described above.

The simple message from this example is that is one should use a detector with pixels as small as possible but with the largest number possible and to place the detector as close as possible to the sample. The other clear consideration is to choose a detector with as much sensitivity to the wavelength of light being used, with a good dynamic range and with as little noise, as possible. The noise and dynamic range play a major role in the accuracy and quality of the reconstructed image.

11.2.3 The Reconstruction Algorithm for Holographic Microscopy

As has been mentioned multiple times in this chapter, the mathematical reconstruction of the image is crucial in DHM. The actual algorithm used depends on the exact optical configuration and the speed of processing balanced against the resolution required of the final reconstructed image. There are a growing number of open source software packages that operate in many computer languages. A recent trend has been to use graphic cards, which have the capability of high speed processing of certain mathematical tasks, including those required for image reconstruction.

Probably the simplest mathematical solution is to consider the optical configuration shown in Figure 11.4, the so-called Fourier holographic method. It should be recalled that the lens in an optical system effectively takes the Fourier transform of the incoming light. (This process is explained in Chapter 12). The basic recording system is shown in 11.4a, where the light coming from the object (either in transmission or reflection) is collected by a lens and directed onto the detector array along with the reference beam, which will also have passed through a lens. Optically the recorded image is then a Fourier transform of the light field emanating from the object. Figure 11.4b then presents optically what is undertaken in a computer. The computer takes the Fourier transform of the hologram to produce three "images". The central one is the reference beam and the +1 and −1 images are the two mirror images of the sample reconstructed. The parameters of the Fourier transform can be altered to change the effective magnification of the computer-generated lens and also the focal point from which it is reconstructing the image. One consideration that does have

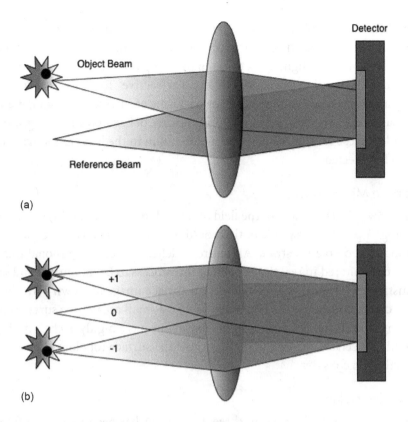

FIGURE 11.4 (a) Image construction, (b) image reconstruction, where the lens is a computer-generated lens to take the Fourier transform of the hologram.

to be made is that of "phase wrapping". In the interpretation of the image it is not possible to determine if a black area in the hologram is caused by a path length difference of $\lambda/2$ or a whole multiple of $\lambda/2$. This is known as phase wrapping as the wave has undergone an entire cycle and reaches round to the same point. Again a number of different methods are used to ensure that this does not happen, with the simplest option being to record two images with a slight change in the optical path for the reference beam.

Other software reconstruction methods are possible but this example illustrates the basic configuration and demonstrates the role of the computer-generated lens in the reconstruction. The computer-generated lens could have infinite optical power but the recorded hologram only has a certain level of information, as described above, due to the number and size of the pixels and hence one cannot have an image with in effect limitless resolution. For further details on the options available for image reconstruction the reader is directed towards the books and review by Kim (2010, 2011) discussed earlier.

11.3 PRACTICAL APPLICATIONS OF DIGITAL HOLOGRAPHIC MICROSCOPY

We now need to consider when DHM may be a suitable method to use for an optical microscopy challenge. The main advantages are that we can record a full three-dimensional image of the sample in a single exposure, though increasing the number of exposures can help to add resolution. The contrast mechanism in the resulting image is provided by differences in the optical path that the light takes through the sample, either due to a different physical distance or due to travelling through a different refractive. This means that the methods of labelling a sample used in many of the other forms of microscopy, using fluorescent or chemical markers or the physical structure, in the case of harmonic imaging, are no longer available unless these changes also lead to an alteration in the local refractive index that is sufficient to be detected.

11.3.1 Surface Microscopy

The initial growth in DHM was in the field of optical metrology and in particular for surface mapping. A practical example of this could be to map out surface etching, or deposition, of a material onto a substrate. An entire widefield, three-dimensional image can be recorded and monitored in real time enabling the process to be monitored and optimized. This is illustrated in Figure 11.5. Here the final surface image is shown with the height of the surface colour coded and the insert shows how a small area of the surface changed its height during an etch and renewal process. Thus if one has a highly backscattering sample, ideally highly reflective, and one wishes to know the surface topology, DHM is probably the first method to consider.

11.3.2 Particle Tracking

The ability to image at high-speed in three dimensions has been used most frequently to track moving particles both biological and mechanical. In the case of biological samples one application has been in the ability to monitor the movement of sperm in order to assess their viability for fertilization (Su, Xue and Ozcan 2012). This system actually used two different colours of LED to obtain the images at high speed and to solve any issues of phase

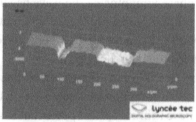

Chemical-etching measured in real-time

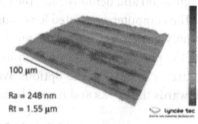

100 μm

Ra = 248 nm
Rt = 1.55 μm

Surface finish measurement

FIGURE 11.5 Digital Holographic images showing the final surface finish of a sample undergoing a chemical etch with the inset demonstrating the height change of an area within the image (Image courtesy of Lyncée Tec SA).

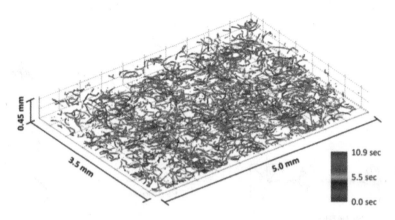

FIGURE 11.6 High-speed three-dimensional tracking of sperm using Digital Holographic Microscopy (Image by permission, Su, 2010, PNAS).

unwrapping described earlier as using two wavelengths of light enables the exact phase to be determined. The resulting movement tracks are shown in Figure 11.6. This example illustrates all of the advantages of the method. The images have enabled high-speed three-dimensional data to be recorded with minimal perturbation to the sample. It is not, generally, possible to add fluorescent markers to sperm if they are subsequently going to be used for fertilization, hence other forms of microscopy are not suitable. Similarly the light intensity should be kept as low as possible to prevent damage to the sample and also to ensure that the light does not perturb the measurement by either attracting or repelling the sperm.

11.3.3 Cell Imaging

The main disadvantage using DHM is that one cannot, in general, label a specific feature of the sample. The method is not really amenable to fluorescence imaging of genetically modified samples expressing a fluorescent protein. However, there are some interesting options where, compared to DIC imaging, the phase changes within the sample can be quantified. This has been applied to red blood cells, for example to study their volume and haemoglobin dynamics (Popescu et al. 2008). This method was then further improved to make clinically relevant measurements on single cells (Mir, Tangella and Popescu 2011).

The label-free nature of the imaging process has been used to advantage in the case of neuronal imaging but also combined with fluorescence (Pavillon et al. 2010). This combination of methods enabled both the physical structure to be quantified along with the dynamic physiological processes taking place within the tissue. The DHM was used to provide real time information on absolute volume, shape and the intracellular refractive index as well as activity through calcium-based fluorescence markers.

11.3.4 Total Internal Reflection Digital Holographic Microscopy

Another more niche application of DHM has been to combine the method with total internal reflection microscopy. The full details of total internal reflection microscopy are covered in Chapter 12 where the focus is on fluorescence excitation. In DHM the sample is not

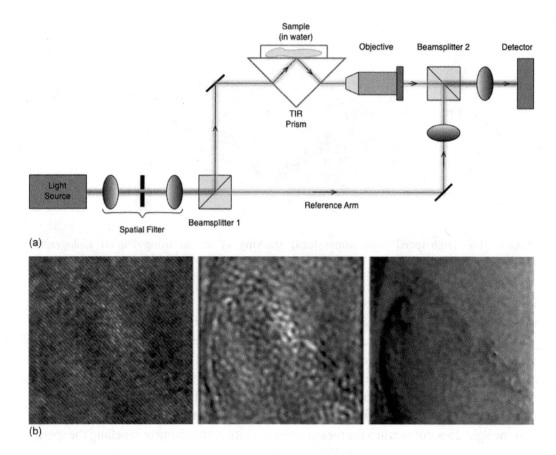

(b)

FIGURE 11.7 (a) Configuration for total internal reflection digital holographic microscopy, (b) image recorded and subsequently reconstructed for TIRDHM (Image by permission Ash 2009, Applied Optics).

fluorescently labeled, but changes in the local refractive index where the sample is in contact with the glass interface can be imaged. The configuration used is shown in Figure 11.7 (Ash III and Kim 2008). The exact principles of total internal reflection microscopy can be found in Chapter 12 but a brief explanation is provided here.

Light is sent onto the sample, which is mounted on a coverslip, or directly on a prism in this configuration, and the light reflects off the interface due to total internal reflection. However, the light field reflected off has a very weak field, which propagates into the sample (i.e. through the glass interface) a short distance of around 50–100 nm. This so-called evanescent wave affects the level of light that is reflected from the interface depending on the local refractive index in contact with the surface. Using the holographic method this local change in refractive index can be measured using the resulting holographic image. One advantage of the holographic method here is that to obtain a wide field of view the light sheet comes in, and out, at an angle, causing distortion on the recording camera. This aberration can be corrected by using the correct form of optic in the computer reconstruction. The method thus provides accurate images of local changes in phase, linked to local refractive index changes, up to a depth of 100 nm into the sample (Figure 11.7b).

11.4 SUMMARY

Digital holographic microscopy is a new method that has developed in the last ten years as computers and camera chips have improved exponentially. It enables high-speed three-dimensional images to be recorded and reconstructed. The core equipment required to record such data is generally low cost and systems are now commercially available, and home building of a system is possible with image reconstruction software being widely available. DHM is emerging as a very popular method for examining surfaces to produce topological maps with ease and is of growing interest in the life sciences community. The real strength for life sciences applications is in the ability to image a sensitive sample without the requirement to add additional labels, in particular fluorescent labels. However, this lack of specific label is also its greatest weakness for some applications. It can be expected that the field will grow rapidly in the next few years as computers and cameras improve further and there is a greater interest in minimally invasive three-dimensional imaging.

REFERENCES

Ash III, W. M. and Myung K. K. 2008. "Digital Holography of Total Internal Reflection". *Optics Express* 16(13): 9811–20.

Ash, W. M., L.G. Krzewina and M.K. Kim. 2009. "Quantitative imaging of cellular adhesion by total internal reflection holographic microscopy". *Applied Optics* 48: H144-H152.

Gabor, D. 1948. "A New Microscopic Principle". *Nature* 161: 777–8.

Gabor, D. 1949. "Microscopy by Reconstructed Wave-Fronts". *Proceedings of the Royal Society A* A197: 449–69.

Hao, Y. and A. Asundi. 2011. "Resolution Analysis of a Digital Holography System". *Applied Optics* 50(2): 183–93.

Isikman, S. O. et al. 2011. "Lens-Free Optical Tomographic Microscope with a Large Imaging Volume on a Chip". *Proceedings of the National Academy of Sciences* 108(18): 7296–7301.

Kelly, D. P. 2009. "Resolution Limits in Practical Digital Holographic Systems". *Optical Engineering* 48(9): 095801.

Kim, M. K. 2010. "Principles and Techniques of Digital Holographic Microscopy". *Journal of Photonics for Energy*: 018005. http://photonicsforenergy.spiedigitallibrary.org/article.aspx?doi=10.1117/6.0000006.

Kim, M. K. 2011. "Digital Holographic Microscopy: Principles, Techniques and Applications". Springer Series in Optics Sciences, New York, ISBN 978-1-4419-7792-2.

Leith, E. N. and J. Upatnieks. 1962. "Reconstructed Wavefronts and Communication Theory". *Journal of the Optical Society of America* 52(10): 1123.

Leith, E. N. and J. Upatnieks. 1963. "Wavefront Reconstruction with Continuous-Tone Objects". *Journal of the Optical Society of America* 53(12): 1377.

Mir, M., K. Tangella and G. Popescu. 2011. "Blood Testing at the Single Cell Level Using Quantitative Phase and Amplitude Microscopy". *Biomedical Optics Express* 2(12): 3259.

Pavillon, N., A. Benke, D. Boss, C. Moratal, J. Kühn, P. Jourdain, C. Depeursinge, et al. 2010. "Cell Morphology and Intracellular Ionic Homeostasis Explored with a Multimodal Approach Combining Epifluorescence and Digital Holographic Microscopy". *Journal of Biophotonics* 3(7): 432–6.

Popescu, G., Y. Park, W. Choi, R. R. Dasari, M. S. Feld and K. Badizadegan. 2008. "Imaging Red Blood Cell Dynamics by Quantitative Phase Microscopy". *Blood Cells, Molecules & Diseases* 41(1): 10–16.

Su, T-W., L. Xue and A. Ozcan. 2012. "High-Throughput Lensfree 3D Tracking of Human Sperms Reveals Rare Statistics of Helical Trajectories". *Proceedings of the National Academy of Sciences* 109(40): 16018–22.

Wu, X. and W. Gao. 2015. "A General Model for Resolution of Digital Holographic Microscopy". *Journal of Microscopy* 260(2): 152–62.

CHAPTER 12

Super Resolution Microscopy

12.1 INTRODUCTION

As stated at the start of the book there has always been a desire to see, and hence study, smaller and smaller features in plants, animals and materials. The original magnifying glasses and then microscopes started to satisfy this desire but greater detail was wanted. When Ernst Abbe (1873, 1881) proved the absolute limit of optical microscopy due to the wave-like nature of light, people continued to look for improvements. This led to the development of electron microscopy in the early 1930s with the first commercial system launched by Siemens in 1938. These instruments use high-energy beams of electrons, which have a very short wavelength, and hence the ability to image with nanometer resolution. However, significant sample preparation is normally required and electron microscopes are not, in general, suitable for use with live biological samples, nor are they compatible with modern genomic techniques, which frequently rely on fluorescence to provide information of activity within the sample.

The interest in trying to develop optical instruments capable of beating the Abbe limit was further driven by the development of the practical confocal microscope, which enabled imaging of three-dimensional, intact, samples. Groups around the world started to explore the challenge of trying to beat the inherent limitation set on optical resolution by the "laws of physics". In the last twenty years this has been achieved partly by using time as an extra parameter in the detection or excitation, secondly by using high performance computing to extract the maximum quantity of information from multiple data sets and thirdly through the use of patterns of light to extract fine detail from camera images or only excite small regions of the sample. In 2014 the Nobel Prize for Chemistry was awarded to Eric Betzig, William Moerner and Stefan Hell "for the development of super-resolution microscopy". This work has led to a huge growth in the field with variations of several common approaches and new acronyms being developed on an almost weekly basis.

Before looking at the methods in greater detail it is worth reconsidering what are the limitations on optical microscopy, which were originally presented in Chapter 2. Light can be considered as either a wave or a particle, but both manifestations lead to a limit in the resolution that can be achieved using an optical instrument. In the case of the photon this can be determined using Heisenberg's uncertainty principle, where the more accurately one knows the position of a particle, the less one knows about its momentum and hence where the particle may have come from becomes less determined (Stelzer and Grill 2000; Padgett 2009). For a wave, this is set by the diffraction effect where for an objective lens with a numerical aperture (NA) and a wavelength of λ one has a resolution limit of $d = \lambda/2NA$ (with minor corrections depending on the exact definition used for resolution). One can neither focus light to a spot smaller than this, nor in the inverse case, determine where the light has come from to better than this resolution. It is methods that defeat this so-called "Abbe limit" that have become known as super-resolution microscopy.

This chapter will present the most commonly used methods for optical imaging, looking at methods in which the light field is limited in extent, or spatially controlled, and also where novel fluorophores are combined with high-speed sensitive cameras and advanced computing to build up images in effect molecule by molecule. The benefits, and potential pitfalls, of the various methods are brought together at the end to guide readers to the most suitable technique for their particular application.

12.2 TOTAL INTERNAL REFLECTION MICROSCOPY

Total internal reflection microscopy is, in some sense, not a full super-resolution technique as it only provides super resolution in the axial direction. However, despite this limitation the method is an important tool (in particular for life scientists) when combined with fluorescence to study activity at a cell membrane or cell movement across a surface. The method was first suggested in 1956 by Ambrose who considered looking at the interface between a cell and a glass prism. This somewhat neglected paper and technique were then revived by Alexrod in 1981 with the addition of fluorescence excitation through total internal reflection to produce the widely used total internal reflection fluorescence (TIRF) microscopy. Once again, as has been a recurring theme in this book, improvements in technology (primarily lasers and cameras in this instance) enabled an earlier concept to be realized for practical applications, permitting events to be recorded and subsequently analyzed.

12.2.1 Principles of Total Internal Reflection Microscopy

The basic concept behind the total internal reflection microscopy is illustrated in Figure 12.1a. Light, typically from a laser (as this provides the most controlled and monochromatically intense beam) is launched into a glass block or slide. The light passes through the glass, which has a high refractive index, to reach the opposite surface. Outside the glass the refractive index is lower. At a certain angle, known as the critical angle, the light does not emerge from the glass block refracted (see Figure 2.3a) but is reflected. The reflected beam then passes on through the block. The reflection at this point is "perfect" and all the light is reflected with none actually passing out of the glass interface. However, by the laws of electromagnetism there is a weak electromagnetic wave that is present on the outside of the glass block. This wave only extends a short distance from the glass interface and falls off rapidly (exponentially) with distance. Under normal circumstances this wave is not detected; However, if there is a fluorophore close to the glass interface, energy is transferred via the evanescent wave. This energy is sufficient to excite the fluorophore, which

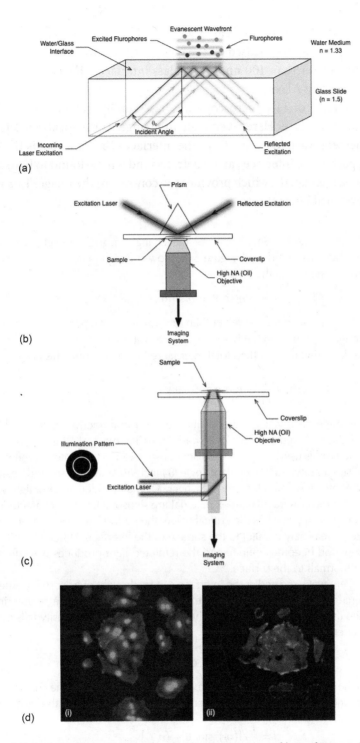

FIGURE 12.1   a) Principle of total internal reflection microscopy; b) total internal reflection via a prism; c) total internal reflection using an objective illumination; d) (i) Epifluorescence image and (ii) TIRF image of the same cell showing the cell adhesion to the coverslip (from microscopy U, by permission of Nikon Instruments).

subsequently emits its fluorescence in all directions. This fluorescence can then be detected using a microscope objective focused on the glass interface. In a typical system the fluorophore needs to be closer than 100 nm to the glass interface. This is the TIRF microscopy technique developed by Alexrod.

The original version of TIR microscopy was based upon local changes in the refractive index at the glass interface. Here, even a small change in the local refractive index can cause the evanescent wave to escape from the interface. The light that is then reflected by the remaining part of the interface can be detected and will be found to have missing parts where the light has "escaped", which provides the contrast in the image. This was the original concept developed by Ambrose.

In Chapter 2, Section 2.2.1, Snell's law for the change in angle as light passes from one medium to another was introduced. Figure 2.3a shows the light passing from one medium (n_i) to a second (n_t) giving rise to the equation

$$n_i \sin\theta_i = n_t \sin\theta_t.$$

The sine of an angle cannot be greater than one so for a given pair of refractive indices the light will no longer be refracted but has to be reflected at a specific point. If in our case $n_i > n_t$ when θ_i exceeds a critical angle then total internal reflection occurs. This angle is defined by

$$\sin\theta_c = \frac{n_t}{n_i} \text{ or more normally } \theta_c = \sin^{-1}\left(\frac{n_t}{n_i}\right)$$

The critical angle clearly depends on the ratio of the refractive indices, which in turn depends on the wavelength of light used and t thus for most TIR microscopy methods a monochromatic laser is used. The change from transmission to total internal reflection occurs without any discontinuity. As the incident angle (θ_i) increases, the transmitted beam becomes weaker and the reflected beam becomes stronger. At small incident angles, light waves passing through to the lower refractive index material are sinusoidal, with a characteristic period. As the incident angle approaches the critical value, the period becomes longer and refracted rays propagate increasingly parallel to the surface of the interface. When the critical angle is achieved, the period becomes infinite and the refracted light produces wave fronts that are perpendicular (normal) to the surface.

The evanescent wave created at the interface can be thought of as partially emerging from the first material into the second and travelling some distance before re-entering the first material. While the evanescent wave is in the second material its intensity falls off exponentially with its intensity being given by

$$I(z) = I_o \exp^{-z/d}$$

where z is the distance into the second material (perpendicular to the surface), d the penetration depth and I_o the intensity at the interface. The penetration depth d is given by

$$d = \frac{I_o}{4\pi}\left((n_i)^2 \sin^2\theta_i - (n_t)^2\right)^{-1/2}.$$

From this it can be seen that by adjusting the incidence angle of the incoming light the depth at which the evanescent wave probes the material above the interface can be adjusted (always assuming the angle is greater than the critical angle).

The important features of TIR microscopy is that the image detected is due to features within the sample that are within around 100 nm of the glass interface. The lateral resolution is determined in the standard way by the numerical aperture of the observation lens ($d \approx 0.61\lambda/\mathrm{NA}$). One thus has a method of recording very thin optical sections, and using a modern digital camera these images can be recorded in excess of 1000 frames per second. It should also be noted that the polarization of the light also has an effect on the penetration depth with "p-polarization" light having a greater penetration depth into the sample.

12.2.2 Practical Implementations of Total Internal Reflection Microscopy

There are two widely used configurations for TIR microscopy; the first using a prism and the second, the one most widely adopted in commercial instruments, utilizing an objective with a high numerical aperture. Figure 12.1b demonstrates the configuration using a prism in an inverted microscope configuration. The light from the laser being used to excite the sample is introduced through a mirror. As the angle of the mirror is adjusted, at the critical angle the light will undergo total internal reflection at the interface between the sample and the glass interface. The objective then collects either the light that leaks out through the sample, where the local refractive index is such that total internal reflection does not occur, or the fluorescence generated by the evanescent wave. The use of fluorescence is now the most widely used contrast mechanism. The collected light is then passed through the remaining optics within the microscope in the conventional manner before being detected by a sensitive camera.

In this system two prism options are possible: the first is with a conventional triangular prism, the second, and more normal, is via a trapezoidal prism. A third option is also possible in which two prisms are used to bring the light into and out of the glass slide. The main complication with this system is that the sample to be investigated needs to be adherent on the glass surface in contact with the prism. This means that the distance from the observation objective to the sample can be sufficiently large that the highest NA objectives have too short a focal length. This reduces the lateral resolution that might be achieved.

The alternative configuration is to send in the light through a high NA objective. A cross-section of this method is shown in Figure 12.1c. The lens in this configuration has a numerical aperture of greater than 1.4 such that the rays most steeply angled are above the critical angle for the glass sample interface. The rear of the objective is then illuminated with an annulus of light, which attempts to enter the sample at an angle greater than the critical angle and is hence totally internally reflected. This illumination causes the evanescent wave to form and the resulting fluorescence is collected through the central part of the objective lens, and passes through a dichromatic mirror before reaching the camera-based detection system.

12.2.3 Practical Considerations for Total Internal Reflection Microscopy

As with many microscopy techniques the first question that has to be considered is the sample to be observed. This normally sets the mounting conditions for the sample and hence the configuration that is most likely to be used. The prism method has greater levels of flexibility and is lower cost (as a specialist objective is not required) but the flexibility does come at the cost of more complex optics and alignment.

The second consideration is then the wavelength of the light source to be used. The majority of TIR microscopy is now fluorescence microscopy so the wavelength of the light must excite the fluorophores of interest. The most widely used illumination for TIRF is a laser due to its monochromatic nature and hence high spectral brightness and also the excellent beam quality making alignment easier. However, there is a disadvantage in that lasers produce "speckle" in the illumination due to their spatial coherent properties, which can lead to non-uniform illumination. This can be partly removed using a simple rotating screen which scatters the beam slightly removing the spatial coherence without disturbing the beam divergence significantly. Due to the narrow spectral bandwidth it is also possible that the laser can cause interference within the sample or glass slide, again leading to uneven illumination.

Alignment of a TIRF system is eased if a standard sample is used before the real sample is placed into the system. One simple sample here is to use a highlighter pen to draw a line on a slide. Most highlighter pens are fluorescent with a yellow one normally being excited by blue light, and an orange pen with green illumination. When the line has dried one is left with a series of fluorescent points on the sample which can be imaged and the angle of illumination adjusted to maximize the fluorescence signal. It is suggested that, if possible, the sample of spots should first be viewed using conventional epi-fluorescence, as should the sample not be seen in this mode the chance of being able to see anything in TIRF is low. The use of epi-illumination also enables the focus to be adjusted to the surface of the sample as there are normal dust particles or minor blemishes on the surface of the slide which provide features upon which to focus.

An alternative to a water-based pen is to use a durable polymer film. A simple way of doing this is through the use of 3,3″ dioctadecylindocarbocyanine (dil formerly from Molecular Probes and now sold by Introvitrogen and Sigma Aldrich). Here the dil should be dissolved in ethanol (around ~0.5 mg/ml) and a single drop placed on a clean slide. Before the solution dries on the surface the slide should be rinsed with distilled water. This will then form a thin sample of fluorescent points that are stable to subsequent water immersion. Fluorescent beads can be used directly in solution but they will tend to "blink" as they drift in and out of the evanescent field making it harder to maximize the detected signal. Once the system has been set up approximately using such test slides the final sample can be introduced and optimization of the final image undertaken.

12.3 STRUCTURED ILLUMINATION MICROSCOPY (SIM)

Structured illumination microscopy (SIM) is another widefield, super-resolution technique, though in comparison to total internal reflection microscopy, SIM produces sub-diffraction limited imaging in two or three axes rather than just the axial direction. SIM does not provide the highest possible resolution, though as a widefield technique it is faster than beam scanned methods. For each optical section that is recorded one has to obtain nine (2D) or fifteen (3D) images, which compares well to localization methods described later, where hundreds of frames are required to produce one super-resolution image. It is probably the most versatile super-resolution method currently available as it can be used on a wide range of samples without requiring special sample preparation, or fluorophores. It is also possible to image at three, or even more, colours simultaneously. The slightly negative aspect is that the method only provides a resolution improvement of around a factor of two, providing a lateral resolution of typically 100 nm and 250 nm axially.

12.3.1 Principles of Structured Illumination Microscopy

As the title of the technique indicates the basic concept behind SIM is the introduction of a controlled pattern of light into the microscope illumination. The core concept behind the method is that the detail within a sample to be imaged is in the fine structure. This consists of features that are spaced only a short distance apart. Figure 12.2a illustrates a sample with a fine linear structure. If this is then illuminated with a fine grid pattern (12.2b) the resulting image is that seen in Figure 12.2c. This consists of moiré fringes, which is the interference pattern between the sample structure and the imposed grid. The fine structure now appears, in effect, as a beat frequency between the imposed frequency and the natural frequency of the sample. The camera records this resulting low frequency pattern in the system. The pattern of light is then moved in a known manner and a further image captured and this method repeated a minimum of nine times for 2D super resolution images (three rotations of the pattern and three translations), or fifteen times for the equivalent 3D image (three rotations and five translations).

The resulting data sets are then manipulated in a computer by taking the Fourier transform of the images and adding the core frequency back in before summing all of the images and taking the inverse transform. Technically the method works as optically a lens takes the Fourier transform of an image. However, due to the numerical aperture of the lens there is a limit to the spatial frequencies (fine structure detail) let through the lens. In SIM

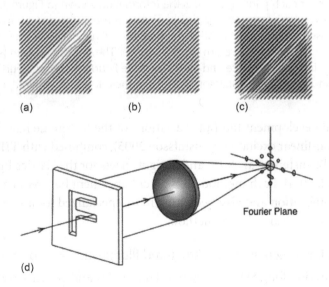

(a)   (b)   (c)

Fourier Plane

(d)

FIGURE 12.2   Structured illumination microscopy: a) sample, b) projected grid, c) high frequency information appears as moiré fringes, d) simple concept of a lens taking a Fourier transform.

this fine structure is translated into a lower "beat" frequency, which does pass through the lens. By taking multiple images with known light patterns, the beat frequency can be reconstructed back into a full image.

As with many techniques it can be hard to assign the first discovery of the method but it is generally attributed to two papers (Gustafsson, Agard and Sedat 1999, 2000). This super-resolution method should not be confused with one of the first uses of patterns of light in microscopy where the pattern was only translated in one direction, three times, to produce an optically sectioned image (Neil, Juskaitis and Wilson 1997). Here the axial resolution was improved but the lateral resolution was at best diffraction limited.

At its heart the SIM method exceeds the diffraction limit by beating the fine structure of an image with a known pattern of light. Figure 12.2d illustrates how a lens takes the Fourier transform of the image. The spatial frequencies for the image are then spread across the optical plane of the instrument. It is the numerical aperture of the lens that determines which spatial frequencies pass through the optical system. With the pattern illuminating the sample the fine structure within the image is pushed into the spatial frequency "pass-band" of the optical system and hence the information can reach the detector.

Figure 12.3 illustrates what is happening in a SIM system in reciprocal space or "k" space (where $k \propto \frac{1}{\lambda}$ or in place of λ a spatial dimension). In a three-dimensional image one has a sphere but for clarity here we have used a two-dimensional representation as a circle Figure 12.3a. The diameter for the circle is set by $k_0' = 2NA/\lambda_{em}$ where NA is the numerical aperture of the objective lens and λ_{em} is the wavelength of the detected light.

If we then illuminate the sample with a known, single frequency, pattern of light which we translate we will have; image with this frequency pattern added, one with it subtracted, and one with the pattern in the middle (Figure 12.3b). If the pattern of light is rotated and then translated in the same manner one will have three images, which lie on a different line in reciprocal space (Figure 12.3c). If the pattern is moved in total nine times (three angles, with three translations for each pattern) one has the information shown in Figure 12.3d. Here the dotted circle shows the information originally passing through the system and any information outside caused by the movement of the patterned illumination is thus additional fine structure on the image when we return to "real" space. The final image is thus made by correctly adding the images in k space and then taking the Fourier transform. Further details can be found in Heintzmann and Gustafsson (2009) and the references within that publication.

Since its initial development multiple variations of the technique have been developed with the use of non-linear excitation (Gustafsson 2005), combined with TIRF for enhanced axial resolution (Brunstein et al. 2013) and in light sheets for the "lattice light sheet microscope" (Chen et al. 2014). Each method adds an enhancement but frequently at the cost of experimental complication and also becoming more specialized for a specific application. At the core SIM is a basically simple method.

12.3.2 Practical Implementations of Structured Illumination Microscopy

The basic configuration for a SIM is shown in Figure 12.4 and systems are normally operated in the epi-configuration. The illumination source may be a laser or an alternative monochromatic source such as a high-power LED or halogen bulb and dichromatic filter.

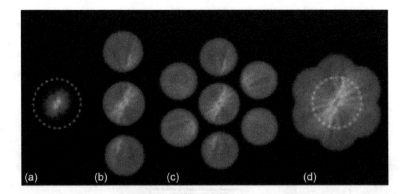

FIGURE 12.3 Structured illumination microscopy in reciprocal space: a) simple image, b) three images with a translation of the spatial illumination, c) three images with a rotation of the illumination d) combination of three rotations and translation of each rotation, with the dotted line showing the normal image illustrating the increase in information passing through the system (Image by permission Heintzmann and Gustafsson 2009, Nature Photonics).

The illumination must be both spatially and temporally consistent. If a laser is used then the spatial coherence has to be removed or there is a risk of speckle appearing in the illumination. The uniform illumination is then passed through a sinusoidal grid of the required line spacing before being directed onto the back aperture of the objective lens and onto the sample. The grid needs to be capable of being both translated and rotated with a precision that is around 1/100th of the line spacing. This ensures that there are no spatial artefacts in the final image.

The resulting signal is then returned back through the system via a dichromatic mirror to the camera. A high-quality camera is required, preferably with 10-bit, or better, uniform dynamic range. The image processing required can be undertaken on a standard personal computer though if high-speed, real-time imaging is required the system requires a significant level of memory and a fast disc for data storage.

Figure 12.4b shows an alternative method of generating the required pattern of light using an interferometric method. The light is directed onto a diffraction grating which produces two symmetrically diffracted orders. These two beams are then recombined at the focus of the microscope objective leading to a striped interference pattern being produced in the sample. Due to the high numerical aperture of the lens the interference pattern only exists for a short axial distance thus providing a level of optical sectioning.

12.3.3 Practical Considerations for Structured Illumination Microscopy

As mentioned above, unlike the other super-resolution methods described in this chapter no special sample preparation is required. The sample can be labelled using conventional fluorophores and then imaged on a conventional microscope with only modified illumination optics to generate the grid within the sample. Although the method provides optical sectioning it is not possible to image deeply within the sample unless it is relatively transparent. If there is scattering within the sample then the grid pattern is lost and the method will no longer provide the required resolution. In order to minimize the loss of the pattern

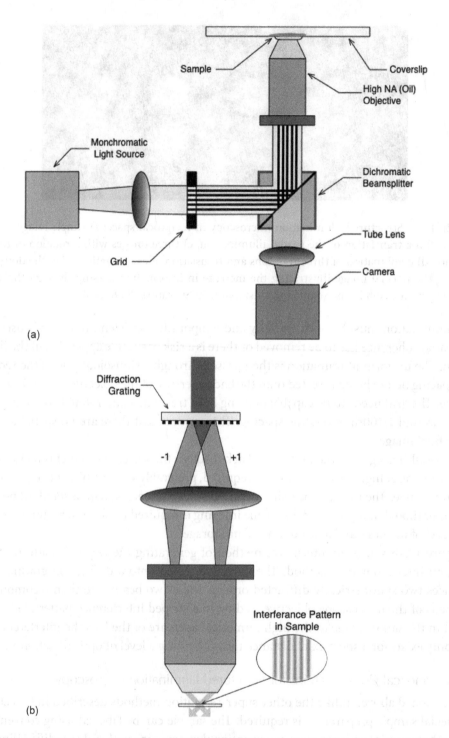

FIGURE 12.4 a) Structured illumination optical configuration with the inset showing a conventional image and a SIM image, b) interferometric method of producing the grid giving improved depth resolution.

before it reaches the sample, care is required to ensure that the correct index matching oil is used with the high NA objective lenses, even taking into account the laboratory temperature to ensure a good match to the glass slide. It is also possible to select slides of the correct thickness by looking at them in reflection in a confocal microscope or even to observe the reflection pattern from a laboratory fluorescent tube when Newton's rings can be seen if the slide is flat enough. Aberrations within the sample can also affect the grid pattern leading to a loss of resolution and also striped artefacts appearing in the images. These can be corrected through the use of adaptive optics though clearly at the expense of a more complex system (Débarre, Booth and Wilson 2007).

12.4 LOCALIZATION MICROSCOPY (STORM/PALM)

The concept of localization microscopy is to excite only a small number of fluorescent molecules within the sample and to detect the emitted fluorescence on a sensitive camera. Software then determines the exact emission point of the fluorescence and this is recorded. This process is repeated multiple times to eventually build up an image, in effect, molecule by molecule. As one can find the centre of a fluorescence spot on a camera to perhaps 1/100th of a pixel the resolution of the system can be around 10 nm. Again, this is a wide-field imaging process but due to the requirement for multiple camera images the method is not fast, though there are some recent advances which are starting to address this issue (Cox, 2012).

Within the field of localization microscopy there are a very large number of acronyms though the two core techniques are STochastic Optical Reconstruction Microscopy (STORM) (Rust et al. 2006) and Photo Activated Localization Microscopy (PALM) (Betzig et al. 2006). PALM is sometimes referred to as FPALM (Hess, Girirajan and Mason 2006) when fluorescence is added at the start but basically the method is the same. Throughout the rest of the chapter we will concentrate on the use of the terms STORM and PALM as these are the two most commonly used methods.

12.4.1 Principles of Localization Microscopy

The core concept of localization microscopy is that one excites only a small number of fluorescent molecules within the sample, such that each emitter can be considered as a single molecule. These single emitting points are then magnified onto a sensitive camera and an image recorded. A second illumination is then placed on the sample and a second image recorded, and this is repeated multiple times. Due to the magnification each emitted point on the camera is spread over multiple pixels, and software is then used to find the centre of mass of the spot, and this point is then taken as the emitting point of the single molecule.

Figure 12.5a illustrates this process on the camera. The centre of mass of the detected light is normally determined using one of two methods. In the first the area around an illuminated set of pixels is subject to a threshold to determine the pixels receiving light, and those not detecting photons. This resulting image (with pixels on or off) is then analyzed and the perimeter of the illuminated pixels fitted to a circle. Subsequently the centre of this circle is found and this position, with sub-pixel accuracy, used to determine exactly the position from which the molecule was emitting photons.

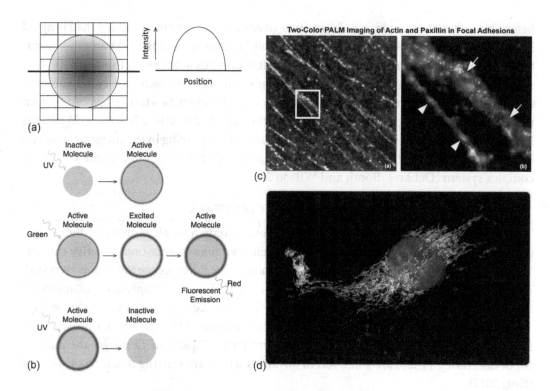

FIGURE 12.5   a) Single molecule emission on multiple camera pixels with an intensity profile, b) the PALM activation, imaging and reactivation cycle, c) PALM image (Image credit Micha Weber/Shutterstock), d) Palm image (courtesy of Zeiss Ltd).

The second method is to draw a series of line profiles through an area of pixels that have been illuminated. The position of the peak intensity is then found for each of these line profiles and the position at which these centre overlaps is used as the position of that molecule. This process is more complex though it leads to a more accurate result. There are multiple minor variations on this method but all have the common concept of curve fitting to an intensity profile.

PALM: This is the more complex method of localization microscopy, generally requiring specialized fluorophores which can be switched into a fluorescent state and then off again. This process is illustrated in Figure 12.5b. The sample is initially labelled with molecules that can change their optical properties when exposed to ultraviolet light. A low level of UV light is then directed onto the sample which "switches on" a few of the molecules. This normally occurs due to a change in shape in the molecule induced by the absorption of a photon of UV light. This then makes the molecule fluorescent when excited with longer wavelength light, typically green or red. With a few molecules now "active" a fluorescence image is recorded using the correct excitation light. A second pulse of UV light is then sent onto the sample, which will deactivate those molecules in the "on" state and activate a set of molecules previously in the off state. A second fluorescence image is then recorded and the cycle repeated multiple times in order to record a full image. The image stack can then be analyzed as described above to build up the image based upon the fluorescent points.

STORM: In the STORM method the sample is illuminated with a lower level of light. This is only sufficient to excite a small number of the fluorescent molecules that will be present in the sample. In each captured image it can be assumed that a random selection (hence the term Stochastic) of the molecules will be excited. By taking multiple images, statistically there will be a bright pixel for each molecule within the sample. STORM can also be undertaken using photo-switchable fluorophores, frequently dyes such as Cy5 where the fluorescent excitation light can also send the molecules into a "non-fluorescent" state, but in which another excitation photon can then reactivate the molecule. This means that only a subset of molecules is excited on each occasion.

12.4.2 Practical Implementations of Localization Microscopy

The basic configuration required for localization microscopy is a fluorescent microscope with a high-quality, sensitive, low-noise camera. The illumination needs to be able to switch wavelength very rapidly and to have its intensity controlled so that the correct number of molecules are fluorescent for any one image. This can be achieved using laser illumination but the latest systems generally use LEDs, as described in Chapter 2. The illumination needs to be uniform across the sample and the use of Köhler illumination (Chapter 3) is required to ensure this. As with all the super-resolution methods the highest possible NA lens should be used, and ideally the psf of the objective should contribute to about 10 pixels on the camera to improve the centre of mass determination. It is also critically important that the microscope structure is stable and free from vibrations, as is the case in all super-resolution methods.

There are now multiple algorithms available for the analysis of the data sets that have been collected using the methods described above. These include open source software and plug-in modules for ImageJ and Fuji. In order to reduce the number of frames that might be required to build up an accurate image, more detailed statistical methods have been adopted to build up the final picture. In particular using a Bayesian approach has led to nearly video rate super-resolution imaging, at least over a small area of the field of view (Cox et al. 2012). As with all such approaches the light level reaching the camera to produce good enough contrast images is frequently the limiting factor in the speed.

12.4.3 Practical Considerations for Localization Microscopy

As stated above, one of the main considerations is to ensure that the microscope and camera are stable and there is no risk of focus drift during an imaging run. Some commercial systems incorporate an active focus stabilization technique in order to minimize such risks. A main consideration in localization microscopy is the use of a reasonably fast computer with high levels of memory as the image files generated and manipulated can become very large, very quickly. Many groups record image sets during the day and then send them to a local computer cluster for processing overnight, in particular when dynamic events are being recorded.

The other main consideration is which fluorophores to use and the list is constantly increasing. Combinations of fluorophores are also now available where an activator fluorophore is linked to a switchable reporter and the activator molecule then "turns on" the switchable reporter. The list of suitable fluorophores is now so large, and developing so quickly that readers should look at the most recent publications within a specific field in order to determine the best current fluorophores for their specific application.

12.5 STIMULATED EMISSION AND DEPLETION (STED) MICROSCOPY

Stimulated emission and depletion (STED) microscopy is a beam scanned technique which requires very accurate timing of ultra-short pulses of laser light, high-speed detectors and suitable samples. The method was first described in 1994 (Hell and Wichmann 1994) and then reported as a practical system five years later (Klar and Hell 1999). The process uses one laser beam to excite a fluorophore and then a second beam, a very short while afterwards, with a "doughnut" intensity. This second beam removes the fluorescence from the molecules except in the centre of the doughnut beam which is black. The second beam is then switched off and the detector activated to collect the fluorescence from the remaining fluorescent molecules. As the centre of the doughnut beam has no light present it cannot diffract and hence can be made, at least in principle, arbitrarily small.

The method has produced images with spatial resolutions in the lateral direction of less than 10 nm but the imaging process is slow and complex. For small fields of view 80 frames per second have been achieved. The light levels reaching the samples are high and thus the STED technique is not suitable for all samples.

12.5.1 Principles of STED

The core concept of STED microscopy is illustrated in Figure 12.6. A diffraction limited spot is initially excited in the sample using a short wavelength laser tuned to the peak of the absorption of the fluorophore. This pulse of excitation is then terminated quickly (typically after around 1 ns) with a number of molecules now in an excited state within the limited excitation volume. A second pulse is directed at the same point in the sample but this light is at the emission wavelength of the fluorophore. When this light reaches the excited molecules it causes them to return to their lower energy level through the process of stimulated emission. This is a very fast and efficient process and so it is possible to remove all the fluorescent molecules from their excited state in an area determined by the shape of the light

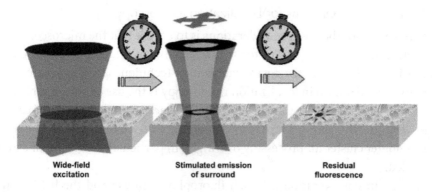

Wide-field
excitation

Stimulated emission
of surround

Residual
fluorescence

FIGURE 12.6 Concept of STED imaging, with the inset showing the energy level diagram for the process.

beam. The crucial feature in STED microscopy is that this depletion beam is shaped with an intensity minimum in the centre, a so-called "doughnut" beam. This means that whilst excited molecules around the outside are returned to their ground state, and can no longer fluoresce, those in the middle receive no stimulating (depletion) light and hence remain in their excited state.

The second, depletion, beam is switched off after around 1 ns so that the molecules left in their excited state have not yet started to emit all their fluorescence due to their relatively long fluorescent lifetime. Once the depletion beam is switched off the fluorescence detector is activated and any photons subsequently emitted from the remaining small volume of excited molecules are detected. The method works due to the highly efficient process of stimulated emission and by selecting a fluorophore with a lifetime sufficiently long that a high percentage of the molecules are still in their excited state after about 2 ns when the detector is activated. However, one does not want a fluorophore with too long a lifetime as one wants the molecules to return to the ground state fairly quickly so the process can be repeated to build up the image.

As the centre of the depletion beam is dark (contains no light) it cannot diffract and hence the centre spot can in theory be made infinitely small. The lateral resolution of the method can be calculated as

$$\Delta r \approx \frac{d_{psf}}{\sqrt{1 + I_{max}/I_s}}$$

where Δr is the lateral resolution, $d_{psf} = \lambda/2NA$ is the diffraction limited psf, I_{max} is the peak STED laser and I_s is the threshold intensity needed to achieve saturation of the stimulated emission (i.e. total depletion of the excited molecules).

12.5.2 Practical Implementations of STED

The basic optical configuration for STED is shown in Figure 12.7 with the left hand side showing the shape of light delivered to the samples by the different beams and the shape of the excited molecules. Two lasers are required, tuned to the excitation and stimulated

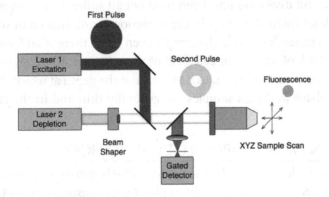

FIGURE 12.7   Instrumentation for STED microscopy.

emission maxima of the fluorophore . Two-photon excitation is possible but typically this is not used as this adds further complexity to an already technically challenging microscope. The first laser is expanded and then passes through the beam scanning system onto the sample. In the initial STED papers the specimen was sample scanned and this is still used in many systems but is inherently slower than beam scanning and not suitable for many specimens.

The first laser is then switched off after around 1 ns, which is typically sufficient time to excite a number of fluorescent molecules within the focal volume of the sample. The second laser is then activated. This beam is again expanded and then passes through a beam shaping optic to produce the desired doughnut beam (or bottle beam for 3D depletion). This can either be a fixed phase plate or a spatial light modulator, which enables more exotic beam profiles to be used. This beam is again active for only around 1 ns but at a very high intensity to remove excited molecules. The gated detector is then activated and the fluorescence recorded before the process is repeated multiple times to scan the entire field of view.

The crucial technical aspects of a STED microscope are the alignment of the laser beams, the stability of the entire system (hence the reasons for sample scanning in many cases) and the timing of the light beams. As light travels approximately 1 foot in a nanosecond (300 mm) careful control of all the beam paths is required to ensure that the pulse timing at the sample is accurate. This also assumes that the electronic "jitter" (error) on the timing electronics and laser activation is of the order of a few picoseconds (Figure 12.7).

12.5.3 Practical Considerations for STED

The major practical considerations for STED microscopy are the technical complexity of the system, the high laser intensity that reaches the sample and the generally slow imaging processes. These challenges should not be overlooked and the requirement for optical imaging, without methods such as localization of the emission, with optical resolution around 10–20 nm, must be very clear before significant money and time are invested in the method. At present there are few operating STED instruments around the world though they have been commercially available for several years. The method is not one to be taken up by the faint-hearted or those without considerable optical expertise.

The other important consideration is in the selection of the fluorophore and sample. A number of different dyes have now been used but all suffer from photobleaching effects due to the high laser intensities used in the depletion beam. Indeed in some samples the laser intensity can cause damage to the sample even when there is no fluorophore present. The imaging method, of any reasonable field of view, is generally slow, and thus not suitable for dynamic samples. One also needs to consider the depth at which the precise beams will retain their shape and thus samples are generally thin and in the case of biological samples, fixed.

12.6 SELECTION OF SUPER-RESOLUTION METHODS

It has become very clear in the last few years that super-resolution microscopy is a technique that is here to stay and likely to play an increasingly important role in general optical microscopy. However, it should be used with a level of caution and is not a panacea for all fine imaging challenges. Generally one is trading off speed of imaging for resolution. There are likely to be higher levels of light on the sample, hence greater risk of damage or perturbation to any delicate system being imaged. There is also increasing use of computer manipulation of the data and thus careful controls should be used to ensure that artefacts are not t being recorded within the imaging system. With all microscopy methods, one should use "the right tool for the right job" and super-resolution microscopy is no different. The table below provides some considerations that should be employed in the selection of a super-resolution method. All methods are now commercially available.

Imaging Method	Advantages	Disadvantages
TIRF	• Technically simple • Relatively low cost • High speed	• High resolution in one axis only • Thin samples only, axial distance 100 nm max
SIM	• Technically not complex • Reasonable speed • No special sample preparation (standard fluorophores)	• Only a doubling of resolution • Not simple image analysis
STORM + PALM	• Selected fluorophores • Resolution ~ 1 nm possible • Technically not complex	• Only tells you where the florescent molecule is • Generally not fast • Not low light levels • Complex computing
STED	• Resolution to 10 nm	• Slow • High light levels • Highly complex system

REFERENCES

Abbe, Ernst. 1873. "Beiträge zur Theorie des Mikroskops und der Mikroskopischen Wahrnehmung". *Archives Microscope Anat* 9: 413–18.

Abbe, Ernst.. 1881. "On the Estimation of Aperture in the Microscope". *Journal of the Royal Microscopical Society* 1(3): 388–423.

Ambrose, E. J. 1956. "A Surface Contact Microscope for the Study of Cell Movements". *Nature* 178: 1194.

Axelrod, D. 1981. "Cell-Substrate Contacts Illuminated by Total Internal Reflection Fluorescence". *Journal of Cell Biology* 89(1): 141–5.

Brunstein, M., K. Wicker, K. Hérault, R. Heintzmann and M. Oheim. 2013. "Full-Field Dual-Color 100-Nm Super-Resolution Imaging Reveals Organization and Dynamics of Mitochondrial and ER Networks". *Optics Express* 21(22): 26162.

Betzig, E., G. H. Patterson, R. Sougrat, O. W. Lindwasser, S. Olenych, J. S. Bonifacino, M. W. Davidson, et al. 2006. "Imaging Intracellular Fluorescent Proteins at Nanometer Resolution". *Science (New York)* 313(5793): 1642–5.

Chen, B.-C., W. R. Legant, K. Wang, L. Shao, D. E. Milkie, M. W. Davidson, C. Janetpoulos, et al. 2014. "Lattice Light-Sheet Microscopy: Imaging Molecules to Embryos at High Spatiotemporal Resolution". *Science* 346(6208): 439.

Cox, S., E. Rosten, J. Monypenny, T. Jovanovic-Talisman, D. T. Burnette, J. Lippincott-Schwartz, G. E. Jones and R. Heintzmann. 2012. "Bayesian Localization Microscopy Reveals Nanoscale Podosome Dynamics". *Nature Methods* 9: 195–200.

Débarre, D., M. J. Booth and T. Wilson. 2007. "Image Based Adaptive Optics through Optimisation of Low Spatial Frequencies". *Optics Express* 15(13): 8176.

Gustafsson, M. G. L. 2000. "Surpassing the Lateral Resolution Limit by a Factor of Two Using Structured Illumination Microscopy". *Journal of Microscopy* 198(2): 82–7.

Gustafsson, M. G. L. 2005. "Nonlinear Structured-Illumination Microscopy: Wide-Field Fluorescence Imaging with Theoretically Unlimited Resolution". *Proceedings of the National Academy of Sciences of the United States of America* 102(37): 13081–6.

Gustafsson, M. G. L., D. A. Agard and J. W. Sedat. 1999. "I5M: 3D Widefield Light Microscopy with Better than 100 Nm Axial Resolution". *Journal of Microscopy* 195(July): 10–16.

Heintzmann, R. and M. G. L. Gustafsson. 2009. "Subdiffraction Resolution in Continuous Samples". *Nature Photonics* 3(7): 362–4.

Hell, S. W. and J. Wichmann. 1994. "Breaking the Diffraction Resolution Limit by Stimulated Emission: Stimulated-Emission-Depletion Fluorescence Microscopy". *Optics Letters* 19(11): 780–2.

Hess, S. T., T. P. K. Girirajan and M. D. Mason. 2006. Ultra-High Resolution Imaging by Fluorescence Photoactivation Localization Microscopy. *Biophysical Journal* 91(11): 4258–72.

Klar, T. A. and S. W. Hell. 1999. "Subdiffraction Resolution in Far-Field Fluorescence Microscopy". *Optics Letters* 24(14): 954–6.

Neil, M. A., R. Juskaitis and T. Wilson. 1997. "Method of Obtaining Optical Sectioning by Using Structured Light in a Conventional Microscope". *Optics Letters* 22(24): 1905–7.

Padgett, M. 2009. "On the Focussing of Light, as Limited by the Uncertainty Principle". *Journal of Modern Optics* 55(18): 3083–9.

Rust, M. J., M. Bates, and X. Zhuang. 2006. "Stochastic Optical Reconstruction Microscopy (STORM) Provides Sub-Diffraction-Limit Image Resolution". *Nature Methods* 3(1): 793–95.

Stelzer, E. H. K. and S. Grill. 2000. "The Uncertainty Principle Applied to Estimate Focal Spot Dimensions". *Optics Communications* 173(January): 51–6.

CHAPTER 13

How to Obtain the Most from Your Data

13.1 INTRODUCTION

As with many of the areas discussed in this book there has been a revolution in the last fifteen years in the use of microscopy images. Technological advances in both imaging capability and local computing power have driven this transformation. The result is that publications, and even internal reports, are now expected to contain detailed and accurate analysis of microscopy images rather than just a high-quality image to illustrate the work. This should lead to greater planning in experimental work with earlier consideration of the data processing as well as the basic challenges of sample preparation and gathering of the data. Correct interpretation and analysis of image-based data is now a core skill of all instrumentation, but one that is, unfortunately, frequently overlooked until large volumes of images have already been gathered. Although the exact method by which images will be analyzed may not be known at the outset, some thought should be given to ensure data sets are recorded at the most suitable resolution, temporally and spatially, for the task in hand.

In addition to numerical image analysis the other aspect of image processing is manipulation of the image to enhance features to help with the interpretation. Any work that is undertaken here must be done with care and the exact processes used clearly documented and reported. Due to the level of software sophistication now available it is possible to bring out what are in truth very minor items within an image, as major features. It is possible to enhance what are in fact artefacts, and therefore these tools should be used with respect, and thought, over what is "appearing" within an image.

As stated earlier, all a digital image consists of is a two-dimensional array of numbers for a single optical section, or a stack of such arrays for a three-dimensional image. The number in the array encodes the intensity for that point, or pixel, and we visualize this as an image when the number is presented either as a specific colour, or grey level. Image manipulation is then a method of statistically analyzing such an array to look for specific patterns or features (e.g. edges). The mathematical functions used may look for associated pixels within the same image or similarities with another image.

This chapter aims to provide a few guidelines and thoughts which will help the microscope user in recording, storing and subsequently processing and displaying data and images. As software is advancing so quickly few detailed examples are presented but the emphasis is on the approach and considerations that should be taken to a range of challenges. Readers are also advised to review the literature most relevant to their field of interest however, *The Image Processing Handbook* does provide an excellent start point for work in this area (Russ and Neal 2017). This will provide an indication of the software packages that people have used, both commercial and shared, and also home-written.

13.2 BASICS OF DATA COLLECTION

The core to all data processing is to record as good an image set as possible before starting to undertake any manipulation. In the previous chapters suggestions on how to record high quality images have been presented and these will not be repeated here. However, the general rules are:

1. Select the correct lens for the field of view and resolution required.

2. Adjust any zoom so that only the field of view required is recorded.

3. Ensure that the settings in the control software are correct (for example if there are places to record the objective lens details make sure these are correct as this may set the scaling factor in images).

4. Ideally record at the Nyquist limit (where the diffraction limit is recorded over a minimum of two pixels).

5. Select the speed of imaging to record any dynamic events required at the Nyquist limit (each time image is recorded at twice the temporal resolution of an event).

6. Maximize the dynamic range (checking black levels and gain using a histogram).

7. Minimize the noise level (ensure gain is not too high, again using a histogram).

8. If using a commercial system save in the proprietary format as this will save the meta-data with each image set.

9. NEVER save images in a lossy compressed format such as MPEG (for movies), JPEG or GIFs. Movies (series of multiple images) are best saved as a series of TIF images if possible.

10. Ensure all the meta-data is recorded. If using non-proprietary software ensure that the data saved includes the objective lens details, the image scaling, filter selection and excitation source. If these are not recorded in the software ensure a separate record is kept.

11. Consider generating a text file or spreadsheet in which each image name is recorded along with any additional information (experimental protocol used for example) and any meta-data not recorded by the system. It is good practice to save this as a file with

the images, so one always has the associated image data available in a format that is simple to search.

12. If using a system that you do not use frequently it is worthwhile taking an image of a grating or structure whose spatial dimensions are known so that the correct scaling can always be applied to an image.

13. Copy the data off the original instrument control computer. Some microscopy centres always save images to a central server so that this happens automatically.

14. Back up the data, or ensure it is backed up by the facility, but even then take responsibility for your images. This is most important as computer storage systems do fail and both time and money will be lost if experiments have to be repeated.

15. Take great care when processing the data that you work with a copy and that there is no risk of overwriting the original file. Personally, I copy the data to a directory in which I then do the manipulation, minimizing the risk of saving on top of the original images.

It may also be worthwhile recording a few images that may not be used for the interpretation of the experiments but which might provide the image that accompanies a report, publication, poster or the highlight of a presentation. Here one might even increase the gain and adjust the black levels to obtain an outstanding visual image. If all of the data for analysis has been recorded it may even be worth risking photobleaching the sample in order to obtain a high-quality image file.

Although data storage is always falling in price, do not record unnecessary images. This may be by recording long streams of images or very large three-dimensional data stacks going beyond anything that will ever be analyzed. One reason for this is that although the storage may be falling in cost, the time to transfer data from a computer to a portable drive, or even over a local network, can be significant. For example, if one has an XYZ stack that is 1024×1024 pixels by 100 images deep as a 12-bit image (not unusual for a confocal image) this is 157 Mbytes of data. If one does this every five minutes for six hours one has 11 Gb. This might take several minutes to copy to a portable hard drive. If one then considers, for example, extended SPIM monitoring a sample over 24 hours, which is not unusual, this number can easily rise to 2.5 Tb or several hours' transfer time both onto a portable drive and then repeated when one copies this onto a local hard drive for analysis.

If these guidelines are followed it does not guarantee that disasters will not take place but the chance of them happening will be kept to a minimum. The one paramount message is that you are responsible for your own very valuable data.

13.3 SOFTWARE CONSIDERATIONS

There are now multiple sources of software available for the analysis of optical microscopy images and the list below only provides a snapshot of the current position. There are three main areas of image analysis software. The first is open source image analysis software that requires no programming expertise. The second is through the use of programming languages, some of which are commercial, but which then require programs to be written to enable the analysis to be undertaken. Finally there is the commercially available software, which has been specifically written for microscopy image analysis. Each of these will now be discussed in outline. The list is dynamic and does not claim to be exhaustive.

13.3.1 Open Source Image Processing Packages

To some extent the field of image processing was altered forever with the release of ImageJ or the associated integrated package Fuji. This is software that was made freely available through the National Institutes of Health in America and is now internationally maintained by users around the world. The software is written in the Java language and is thus platform independent. As well as the core tools for undertaking basic image processing, users have written "plug-in" modules which undertake specific tasks, or automate a series of image processing steps. The software is easy to install, intuitive and capable of handling all common image formats, including translations from manufacturers' proprietary formats. It is suggested that all computers which may be used for looking at any microscopy images should have ImageJ installed. Fuji is a bundled package using ImageJ as the processing engine but incorporates multiple processing plug-ins. All the example images provided in this chapter were undertaken using ImageJ version 1.47 running on a MacBook Air with 8 Gb of RAM.

Many of the developers of ImageJ modules are members of the Open Microscopy Environment (OME) community. This is a consortium of universities, research laboratories, industry and developers producing open source software and format standards for microscopy data. In particular they have focused on the development of OMERO. OMERO is an open source client-server software package for managing, visualizing and analyzing microscopy images and associated meta-data. Many microscopy centres use versions of OMERO as their main system of storing data from a range of commercial microscopes that typically reside in such institution centres.

A further major open source package has been released by the Broad Institute and is known as Cellprofiler. The Broad Institute is a research institution combining the expertise of Harvard, MIT and the Harvard associated medical institutes. The focus is on advancing the life sciences through multidisciplinary research with a strong focus on using physical science data analysis methods for the life and medical sciences. This package contains a selection of image processing tools, more closely linked to data analysis than ImageJ.

13.3.2 Programming Languages for Image Processing

All computer languages can clearly be utilized to undertake image analysis but there has become a tendency to focus on the three examples discussed below. Many commercial and older image analysis packages have been written in languages such as C++ and Fortran but these require a high level of programming expertise. Those listed below are easier to approach even for people who are not programming experts. It should be appreciated though, that some experience of programming is generally required to undertake data analysis using these fairly low level languages.

Probably the most widely used language in optical microscopy image analysis is MATLAB. The language is only available as commercial software but this includes several packages dedicated to image analysis. In addition there are multiple users around the world, both in academia and commercial research organizations, who make their programs available to all though a program exchange. However, these are normally offered with no support and thus some level of MATLAB understanding is required to operate these programs effectively. MATLAB programs can be optimized to operate very quickly and it is the language most widely used by image processing experts to initially develop methods before they may be rewritten into a very high-speed language. Using the commercial modules that are available it is possible to rapidly bring together multiple tasks to generate a pipeline for image processing. A further important feature is the ability to take the output from an image analysis method and to then process the resulting data in complex ways within MATLAB, or to export numbers to a further package for statistical analysis.

The next commonly used language is Python and its associated modules Numpy and Scipy. This is a much lower level language but is free and is now the language taught to many undergraduates in the physical sciences. There is some similarity in syntax to MATLAB and Python is very powerful at array analysis. Many basic image analysis tools are now available by downloading from a large number of internet sites, though again some programming knowledge is required to enable these to actually function on your images. Python is, however, an easy language to learn and thus the entry level to developing your own processing and analysis tools for a specific application is not too high. Indeed many expert users of optical microscopes are teaming up with experienced programmers from other disciplines to produce dedicated code for specific applications. Python will run on all computer platforms and the code can be optimized to make best use of advanced computing techniques including parallel process threading within the processor. One word of warning: be aware that Python versions 2 and 3 have slightly different syntax and thus when downloading a program ensure that you have the correct version of Python.

The other language used for image processing is the Interactive Data Language (IDL). This is a commercial language specifically designed for image and data processing with a focused package known as ENVI. IDL again requires programming expertise to enable any analysis to be undertaken but many tasks are automated by a single command line. It is less widely taught as a language, due to its specialism, but is widely used in fields such as astronomy. Many of the programs developed in this field can be modified for use in optical microscopy. If a project is planned and collaborators are available from within the astronomy community then this is a route to consider.

13.3.3 Commercial Image Processing Packages

Commercial software can be viewed as being expensive but it is worthwhile considering that if it has taken eighteen months to generate a suitable transgenic line, money spent on software may not seem as large. In particular, if the software is used by a collective of users within an institution the financial burden can be shared. The real challenge is that although many commercial packages will undertake most of the basic tasks required, some tend to specialize in one area, hence several purchases may be required to acquire a powerful and versatile software suite. It should be noted that the comments below are brief and imply no endorsement by the author. The only way to determine which package may be suitable for a specific imaging application is to download a trial package and "play" with one's own data set. Most companies are very happy for this to be undertaken and may even help the potential purchaser with the trial analysis.

The first area to look for image analysis using commercial software is that provided by the microscope companies themselves. The important feature here is that they will be dedicated to using the meta-data attached to each image file and thus this information is not lost during the processing steps. In addition there should not be any compatibility issues, which can frustrate one when using other packages. As digital cameras are used increasingly for optical microscopy, in particular for localization-based super-resolution techniques, camera companies are frequently supplying image processing software specifically designed to control the camera, record the data and then to produce processed images. Thus the equipment suppliers' own software is not to be overlooked as a useful starting point for any image analysis that might be undertaken.

One widely used advanced image analysis package is Metamorph. This software has grown and is now capable of operating some camera-based microscopy systems. One of the major features in Metamorph is the analysis and quantification of fluorescence images. This may be for looking at positional, or temporal, changes between images, or for the co-localization of different fluorophores. It also has excellent features for tidying images to remove content that the user is certain is an artefact. Clearly this capability should be used with great caution, and, crucially, carefully documented, as will be described later.

Another popular package is Imaris sold by Bitplane. Although capable of undertaking multiple forms of image analysis it is becoming increasingly focused on the analysis of big data sets. Traditionally it has also been strong in temporal analysis of image sets recorded over time which has led to the ability to deal with large image sets. This is very useful for image analysis on datasets produced from in-depth multiphoton microscopy and in particular for SPIM-based images.

The final general package to be mentioned is Velocity from PerkinElmer. This specializes in tracking particles, or features, in three dimensions over extended periods of time. This may be groups of cells, or individual components within a cell, and the features may even evolve with time. The software is also well suited to presenting this movement, or general temporal changes, as a series for graphs or vector diagrams showing both the velocity and direction of travel.

There are other packages available which specialize in specific imaging tasks such as Huygens software which is designed specifically for the deconvolution of images. However, it should be noted that all the above packages have some ability in this area. There is also a growth in cloud-based image analysis software such as Oculyze and it will be interesting to see how these succeed. It is possible that the sheer size of some data sets may make this a route to consider so that the cost and complexity of local storage is removed.

13.4 BASICS OF DATA PROCESSING

The core aims of image processing are to:

1. Assist the viewer in observing or communicating information in images.

2. Reduce the effect of human bias or desire to see certain features.

3. Introduce a level of quantification and statistical rigour to image analysis.

As indicated in point 2, there is also a golden rule that in doing any manipulation the operator must not introduce bias. It is exceedingly easy to introduce bias just through the use of the wrong image processing tool, or by running through a processing sequence which introduces bias in an early stage and then indirectly measures this later in the process. A rule is, think what an image processing step actually does. Do not just use it so that the image looks "nice" or so that the features you thought were there now appear. It is very easy to believe what a computer tells you, but remember, in general, it is only doing what you told it to do. The exception is in the use of artificial intelligence in image processing, an area which is evolving very rapidly but which is beyond the scope of this book.

13.4.1 Core Techniques

Figure 13.1a illustrates a standard 8-bit monochrome image of Durham Cathedral. A non-microscopy image has been selected as within this single image all the features that one wishes to discuss are present. In Figure 13.1b an area of the image has been expanded to show the individual pixel values. It is these pixel values that will be operated upon by the various computation operations on the image.

13.4.1.1 Reducing Noise

Noise in an image is a feature that appears that is not part of the actual "scene" on the microscope stage. The noise may have come from the light source, the detector, the digitizing electronics or from a random cosmic ray. Noise is classified in three groups:

Shot noise: This appears as random bright or dark pixels across the entire image. They are likely to be randomly positioned but can appear in a row or column due to a systematic error in a camera or scanning system. Statistically this noise will be Poissionian in nature.

Speckle variations: These again appear as random fluctuations and are caused by the detector or electronics and potentially the illumination source in a scanned system. Generally Gaussian in nature.

Interference or periodic variations: This noise will be banded or appear in specific areas and may be due to external lights or the detector picking up monitors. It has a repetitive frequency due to the scanning of the detector or the light source.

The first two types of noise are random and hence can be removed, or at least decreased in the image, using averaging methods. This might be by taking three identical images and

FIGURE 13.1   a) A monochrome image of Durham Cathedral; b) indicative pixel values for an area of the image.

determining the average for each pixel point, or by using a neighbour averaging process. Here the pixels surrounding each individual pixel are used to produce an average for the central pixel. This helps to remove high individual pixel value. Thus one is using spatial averaging to remove the noise.

As periodic noise appears at specific spatial frequencies in the image such noise is best removed using a Fourier frequency method. Here the image is converted into spatial frequencies, a specific frequency (the noise frequency) is then removed and the remaining spatial frequencies converted back into real spatial information.

13.4.1.2 Uneven Illumination

Although every effort is made to produce an even illumination on the sample, and uniform collection efficiency, this sometimes does not happen. Removing this effect is important as a first stage prior to more detailed image analysis.

Removing measured background: The best way of undertaking this process is to record a blank image with the sample removed. In the case of fluorescence this might be a plastic fluorescent sample or even a slide containing a "sea" of fluorophore. In the case of conventional widefield transmission images this can be achieved using a homogeneous scattering sample. This could be a ground glass screen or a scattering solution (where milk in water works very well). One now has the choice of either directly subtracting the background or undertaking a division of the images. Most modern detectors are now linear in their response to light, and division of the main image by the background image is the best solution to levelling out the background.

Removing interpolated background: It is not always possible to take a background image without the sample and in this case users may select multiple areas across the image, that they are sure are background, to act as a "map". The computer can then measure the intensity at these points/areas and join them up using a polynomial fit. Typically a quadratic (second order) or cubic (third order) polynomial is used to provide the correct compromise between an accurate fit and a sufficient number of data points. One now has a computer-generated background that can be used to correct the original images. Clearly this option has some faults as frequently the selected areas are in the periphery of the image whereas the centre normally has the features of interest. The other complication is that users have made some selection on what they think is the correct background region and hence bias can be introduced.

13.4.1.3 Increasing Contrast

Having removed the noise and then corrected the background the next task is to look at increasing the contrast or dynamic range of the image. This is typically a standard button on image analysis software and is frequently used too early in the process. One should try to undertake the improvements suggested above before adjusting the contrast. It is useful here to look at the image histogram, either of the entire image or a selected area. The histogram plots the number of pixels of a given intensity arranging the pixel values in ascending order. In a colour image three histograms are normally available; one each for the blue, green and red image channels. Figure 13.2a illustrates the combined histogram (black and white) and the separate colour histograms for our standard image.

The contrast and brightness of the image can then be adjusted to obtain the best-looking image, or the most suitable for subsequent image processing. In making manual adjustments the cut-off of the displayed pixels is illustrated in the inset in Figure 13.2b,c. Here the contrast has been decreased (b), and increased (c), significantly to show the effect. In the decreasing image (13.2b) the white pixel values are displayed as a linear line of contrast but instead of intercepting at zero the intercept is now at –205, introducing what appears as a grey haze across the image (in effect all pixels below a certain value are set to a default grey level). In 13.2c the enhanced contrast has now taken all pixels less than 80 and made them black, and all pixels above 174 as white. Figure 13.2d shows the computer equalization of the histogram with a maximum of 0.4% pixels saturated and the histogram extended over the full range. The clouds now become more visible but details have been lost in the trees when compared to the original monochrome image Figure 13.2e. If

FIGURE 13.2 a) A colour image with a combined histogram and one for each colour; b) monochrome image with decreased contrast; c) increased contrast; d) automatically increased contrast using a 0.4% saturation and normalization across the image; e) original image in monochrome; f) inverted image.

subsequent segmentation and thresholding of an image is going to be undertaken then expanding the contrast can make this process easier. However, the main aim is to improve the visibility (to the observer) of features within an image. In this context it is worth remembering that human vision detects contrast as the percentage difference between brightness, not the absolute difference. Thus between a pixel value of 10 and 20 the eye and brain will visualize a difference of 100% whereas 200 to 210 is a difference of only 5% and thus much harder to perceive. This effect can clearly be seen in Figure 13.2f where the only change in the image from 13.2e is that the image has been inverted. If one looks at the clouds the slight variations are much more obvious but the detail in the cathedral towers or trees has been lost.

13.4.1.4 Enhancing Perceived Detail

The methods above have looked at improving the overall image quality. The following are specifically focused on improving the features seen by the observer. The suggestions below can be used to enhance individual aspects of an image to bring out specific features, though this might be to the detriment of the rest of the image. Figure 13.3 illustrates some of the methods mentioned in the text.

Local contrast enhancement: Changes in contrast described above apply equally to all pixels within an image. However, one can locally enhance individual pixels compared to the neighbouring ones. This reduces the effect of pixels that are some distance from the feature of interest. A multiplying table changes a small area of pixels. Typically this is the central pixel and the eight surrounding pixels. For example the central pixel may be multiplied by a number and then combined with the addition of all the local pixels to form one new value. The process is then applied to the next pixel and so on throughout the image.

Local sharpening: A similar process can be applied but here the centre pixel may be multiplied by a number and then the other pixels subtracted to give a sharper edge to the image. This is also known as Laplacian sharpening and is illustrated in Figure 13.3b.

Local contrast reduction, unsharpening mask or Gaussian blur: This is in effect the inverse of the above procedure where a value of the local pixels is added to the central pixel. This is illustrated in Figure 13.3c. An alternative here is to add a Gaussian blur to each pixel by convolving the pixel with a Gaussian intensity profile.

Shadow filter: This adds values of pixels to one side of the central point and removes values the other side, enhancing the edge of features where there is a local difference in contrast. Figure 13.3d.

Threshold filter: Here a pixel value is set and then everything above this value is made bright (255 in an 8-bit image) and everything less than the value changed to 1, or black. TA dual intensity image is produced and this is a very powerful technique for subsequently

a) Natural Image   b) Sharpened Image   c) Smoothed Image

d) Shadowed Image   e) Thresholded Image   f) Enhanced Edge Image

FIGURE 13.3   a) Natural image; b) sharpened image; c) smoothed image; d) shadowed image; e) thresholded image; f) edge enhanced image.

finding the area of certain bright, or dull features, as a simple pixel count is now all that is required. Figure 13.3f.

Edge enhancement: This looks at the local brightness and calculates the gradient of the change in brightness which is then multiplied onto the pixel. This has the effect of finding edges as shown in Figure 13.3f.

13.4.1.5 Monochrome Look-Up Tables and the Addition of Colour

As many publications and reports are now available via the internet, and then normally viewed on a computer screen, the addition of a colour look-up table can significantly enhance features or bias the perception of an image. In a colour look-up table a range of pixel intensities is assigned a specific colour, whereas in a monochrome image it is a grey scale that is assigned to a pixel value. This is not therefore the colour seen in the image but an intensity match of colour. Figure 13.4 shows an image of some mitochondria labelled with the fluorescent dye TMRE. Figure 13.4a shows the raw image and then in 13.4b a non-linear (logarithmic) look-up table has been used, which has the effect of compressing the dynamic range. This does mean that sometimes very faint features now appear from the background, but some detail is lost. Figure 13.4c uses the function of enhancing local contrast but the global contrast has been lost. However, this image might now be easier to threshold in order to select the features of interest. 13.4d has a non-linear colour look-up table and 13.4e a discrete colour map where specific ranges of pixels have a specific colour. It is left to the reader to comment on the perception one receives from the different renditions of the image.

Care must be used in such tables as Figure 13.4f can be confusing, for several reasons. If this image is displayed at a microscopy talk on image processing and one asks what it shows, a common response is that it is an image of a brain or perhaps sometimes a cauliflower. This is partly determined by the look-up table used. In fact this is a topological map with height encoded in colour for the rivers and hills around Durham, UK. This particular colour table is frequently applied in the life sciences and hence one's mind is led to fit the image to a biological interpretation. This image is also missing one other important feature and that is a figure for the scale bar. This has been deliberately missed off the caption. The bar actually represents 20 km and I am indebted to Dr Chris Saunter at the University of Durham for this image.

13.5 PRODUCING QUANTIFIED DATA FROM IMAGES

Having briefly outlined and discussed some of the visual changes that can be made to an image, the remaining sections of this chapter will look at the quantification approaches that can be applied. One important point to remember, throughout any quantification of images, is that some consideration should be given to errors in the numbers that will be produced. Exactly how the error is calculated and subsequently displayed will depend on the exact applications but it is very important that some thought is given in this area. Without any comment on the level of errors from a data set any conclusions drawn have to be questioned. As the number of options in quantification of images is very large the notes below are aimed at providing guidelines and series of ideas that should be considered in the quantification process.

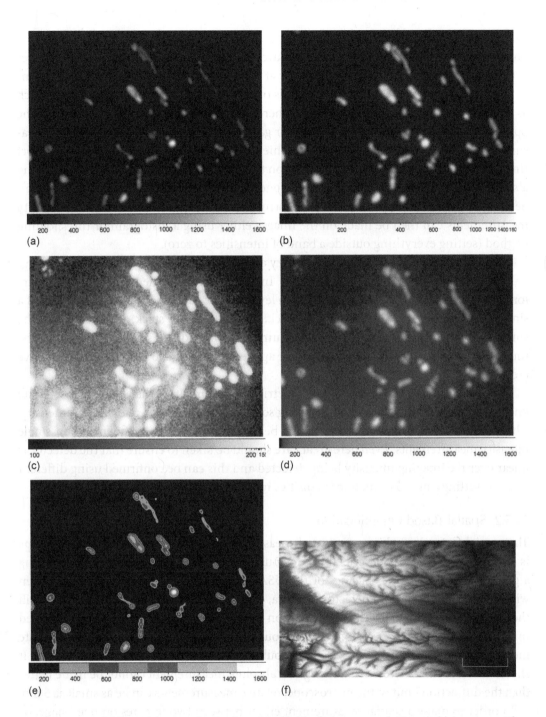

FIGURE 13.4   a) Normal image of mitochondria (field of view around 40 μm); b) image a with a logarithmic monochrome look-up table; c) image a with contrast enhancement; d) image a with a non-linear colour look-up table; e) image a with a discrete colour look-up table; f) incorrect use of a look-up table, see main text for explanation. (Images credit to Dr Chris Saunter, Durham University.)

13.5.1 Intensity-Based Quantification

The first comment in relation to direct measurements of intensity from microscopy images is basically, do not do it if at all possible, or at least try not to make direct intensity comparisons. An absolute measurement depends on many poorly controlled parameters, such as intensity fluctuations, the actual light gathered at that point, dirt or other changes in the optical systems, as well as the exact detector gain, black level and linearity. If direct measurements are made the best that can be achieved is to try to record all images in a short time period, trying not to alter any settings on the instrument. It is also worthwhile having a standard sample (for example a known concentration of fluorescent solution) that can be imaged at repeated intervals to try to normalize the images recorded. Direct intensity measurements can then be made on the images either using a histogram or thresholding method (setting everything outside a band of intensities to zero).

It should also be remembered that the very act of recording an image can affect the signal due to the photobleaching of the sample. In taking a three-dimensional confocal scan, for example, the sample could receive a high level of light. Here it can be useful to record a single section at the start of the imaging stack, image the entire stack, and then repeat the single slice measurement. It can then be assumed that any loss of signal is linear with the image number and a correction factor can be applied to each image slice as described above (Section 13.4.1 on uneven illumination).

A preferred method is to take a ratio-metric measurement, perhaps having a separate known standard within the sample, or using some other feature as a reference of intensity. Changes relative to this standard can then be made with greater confidence than single intensity measurements. Even here great care should be taken to ensure that the detector is linear over the imaging intensity being detected and this can be confirmed using different detector settings and illumination intensities in a test sample.

13.5.2 Spatial-Based Quantification

The crucial factor in making any spatial measurements is to ensure that the microscope is correctly calibrated. As described in previous chapters this can easily be achieved using a grating with known line spacings or the USAF test target. All subsequent images taken with that objective lens and detector (camera, or point detector in a scanned system) will then be correctly calibrated. The one exception to this is when one is using a beam scanned microscope system and one "zooms in". Although the scaling should be correct if accurate measurements are to be made a scale bar should be used to check the exact calibrations. It should also be remembered that although an absolute measurement cannot be made better than the diffraction limit of the microscope, relative measurements can be as small as 5 nm.

In order to make a spatial measurement either between two features on one image, or between the same feature in two different images; one needs to know where the centre, or edge, of the feature is. There are several methods that can be considered in determining this point and these are briefly listed below:

Centre of mass: This is the method used in localization microscopy (Chapter 12 Section 4.1). The intensity of the feature of interest is spread over several pixels and the centre of the intensity patch can be found to up to 100th of the pixel size, and when the magnification of the microscope is included then the resolution can be as great as 5 nm.

Segmentation and centring: Here the perimeter of a feature is found and then the centre of the feature determined. This is based upon the geometry of the perimeter rather than the intensity profile. Thus, the image is frequently first segmented and then the feature circumference found, before the geometric centre determined.

Intersection of lines: A variation on the above two methods whereby two lines, at 90° to each other, are drawn through a feature of interest at the maximum width of the feature. The intersection point is then determined to be the centre of the feature and thus distances can be measured from this point.

The other method, though clearly prone to human bias, is to simply draw a line on the image and to count the length of the line in pixels. This can subsequently be converted into a physical measurement using the scaling of the image. This is clearly not the preferred method to use but for a quick inspection of an image it can provide insights which can subsequently be confirmed using a more automated method.

In three-dimensional data sets if the image slices are arranged as a stack with the depth profile extending through the stack, distances can be measured through the stack using any of the software packages described earlier. There is one warning that needs to be considered here and this relates to the actual distance recorded. As light travels through the sample it is refracted and thus if a sample moves 1 μm in air the sample this will be modified by the refractive index of the surrounding material. If the sample was in water ($n = 1.33$) the focal spot, and hence new image plane, will have moved by 1.33 μm. Many microscopes incorporate this movement allowance into the system and care should be taken over this measurement.

13.5.3 Temporal-Based Quantification

The most important aspect in making temporal measurements is to ensure that the image plane has not moved either laterally, or through the optical axis, as this will cause motion artefacts in the image. Thus great care should be taken to ensure that the sample does not move and if three-dimensional stacks are being recorded that these are checked to ensure there has been no lateral movement as one goes deeper into the sample. One can then use the computer clock as the master timing for the system as all images will be recorded with a very accurate time stamp. The subsequent analysis of the images may then plot a change in position or intensity over time to produce a so-called "kymograph". This is a two-dimensional image in which one axis is time and the other movement. This is a particularly powerful method to monitor a moving feature/organelle within a cell and to provide some quantification of its motion. Its movement in distance can be plotted in the x-axis and then the y-axis is used as the plot of time.

An alternative route is to plot the movement on a standard two-dimensional image with the speed of movement encoded in colour. This can be seen in Figure 13.5 (Chalmers et al. 2012) where mitochondria have been tracked in both native and cultured cells and the

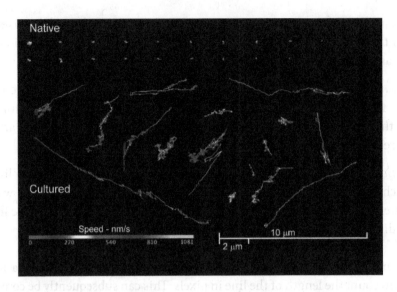

FIGURE 13.5 Tracking mitochondrial movement. (From Chalmers, Susan et al. 2012. *Arteriosclerosis, Thrombosis, and Vascular Biology* 32(12): 3000–11.)

differences in movement displayed as colour encoded tracks with the length linked to the spatial movement, and the colour linked to the speed.

13.6 DECONVOLUTION

No book on optical microscopy would be complete without a comment on deconvolution. This is a computational process in which the optical properties of the system are considered as artefacts within the image and removed. The method is most applicable to images that have a third dimension but nearly all images obtained on an optical microscope can be deconvolved. Essentially it removes the blurring present due to the finite aperture of the objective lens, though to some extent all the optical components do add blur. In particular it can be used to obtain some level of optical sectioning from widefield microscopy, but the method can also be used to improve images from a confocal system.

To appreciate how deconvolution operates one needs to briefly reconsider how a microscope image is formed, in particular linking back to Chapter 2 Section 2.2. If one considers an infinitely small source of light within the sample (for example a single molecule) the actual pattern of light that is detected by the microscope is a combination of this point and the point spread function (psf) of the optical system. This is dominated by the diffraction of light by the objective lens, though other optical elements all add slightly to the actual spot that is seen. Technically this is known as the convolution (mathematically shown as ⊗) of the psf with the feature in the sample.

If we consider for a moment light being focused by the optical system, the actual focal volume produced will look something like that shown in Figure 13.6a. In the plane of the sample one has a disc of light that consists of a series of concentric rings known as the Airy pattern (Chapter 2 Section 3.2). Along the optical axis (z in the diagram) one has an elliptical pattern with the overall effect being a region shaped like a rugby ball, or American

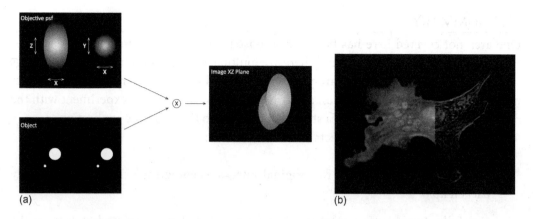

FIGURE 13.6   a) The effect of the point spread function on an image point, b) an image showing on the left an area not deconvolved and the right with the out of focus light removed by deconvolution (note this is only a thin cell). (Image by permission of Scientific Volume Imaging BV.)

Football. If we then place a point source at the focus of the microscope the optical system would in effect see this as coming from anywhere within the "rugby ball" volume. If the molecule was towards the edge of the volume its light would still be detected as if coming from the same volume in space, though more faintly than if it was at the centre. The third position for the point of light is outside the elliptical volume (or the focal plane) and again some of the photons would pass through the optical system and be detected. Deconvolution in effect helps to position the source of light more accurately by mathematically removing the optical function of the lens and assigning the photons collected to the correct position in space.

Thus, deconvolution does have a similar role to the pinhole in a confocal microscope helping to remove the out of focus blur. However, deconvolution can also be applied to confocal images further helping to enhance the image quality. The process needs to be applied to each pixel in the entire image stack and thus a high performance computer is required. Figure 13.6b illustrates an image taken from a single cell where deconvolution has been applied to half of the image. The removal of out of focus blur is clearly enhancing features within the image and increasing the contrast.

There are now multiple sources of software available for undertaking deconvolution including packages supplied with optical sectioning microscopes. They operate in two main modes: one in which the optical psf is assumed to be known (or measured), the second in which the psf is mathematically determined using the known properties of the system (e.g. lens NA) and information from the image stack in a process known as "blind deconvolution". Both methods produce improvements in the image quality though conceptually using a measured psf for the microscope seems intuitively to be a better route to follow as one is starting with some known information.

If deconvolution is going to be used one should always improve the image quality using the methods described above, before the deconvolution algorithms are applied. Noise within the image will significantly affect the performance of any deconvolution, as bright spots may be reassigned to the wrong position in the final image.

13.7 SUMMARY

One area not covered here has been the reconstruction of three-dimensional data sets. Here the field is very wide open and there are multiple variations on how such renderings may be made. Many of the software packages described above have a range of reconstruction algorithms and this is again an area where it is best that readers experiment with the different options to produce a data stack that is suited to their specific application.

The three most important aspects to this chapter are:

- Only manipulate copies of the original images, keeping the original images safely backed up.

- Carefully document and report all operations that have been undertaken on the images, so that any process can be repeated.

- Finally, in any quantification undertaken consider the errors in the figures obtained. Many of the software packages do not consider the errors from any image analysis and this is thus left to the user and it should not be forgotten.

In summary, once high-quality image data sets have been taken, allow time to process the results. It may very well be that a number of approaches will need to be applied to each data set before the optimal route to presentation and quantification is discovered. Ideally a test data set should be taken and then experimented with, as it may be that the exact imaging procedure may need some slight changes to ensure accurate and reliable data processing. It is always much better to discover this before recording large volumes of data!

REFERENCES

Chalmers, S., C. Saunter, C. Wilson, P. Coats, J. M. Girkin and J. G. McCarron. 2012. "Mitochondrial Motility and Vascular Smooth Muscle Proliferation". *Arteriosclerosis, Thrombosis, and Vascular Biology* 32(12): 3000–11.

Russ, J. C. and F. B. Neal. 2017. *The Image Processing Handbook*. 7th edn. CRC Press.

CHAPTER 14

Selection Criteria for Optical Microscopy

14.1 INTRODUCTION

This chapter provides some guidelines on selecting which optical microscopy method is best suited for a specific application. The preceding chapters have described systems that produce data sets ranging from single transmission images through to complete three-dimensional data sets recorded over extended time periods. The images may contain fluorescence lifetime derived information on the local environment as well as chemically specific mapping for each voxel in an image stack. Financially they range over a thousand times in cost (three orders of magnitude), indeed if one includes the latest generation of "webcam" microscopes this becomes over ten thousand times. These costs are even without the extra investment that may be required in a dedicated laboratory area and associated data support infrastructure. Thus making a commitment in optical microscopy is an important decision.

There have been multiple cases of researchers who felt they needed the most modern, sophisticated instrumentation when in fact their task required a robust, reliable and much more basic system. When new methods have been introduced in the past there has been a scramble to obtain funding for the new "toy" that will "transform my research", and frequently people have subsequently felt let down when the expected revolution does not occur. A classic example occurred with the introduction of multiphoton microscopy in the mid to late 1990s. When these systems initially appeared, as commercial instruments, the laser sources required significant expertise to operate (11 manual knobs to be adjusted!) and the microscopes tended to demand regular realignment. The perception of potential users was that such microscopes were the next generation of three-dimensional imaging systems, supplanting the confocal microscope. New users were then sometimes disappointed when the results they obtained were not as good as with their previous confocal, with lower resolution and slower imaging speeds. Both of these statements were true as the longer excitation wavelength leads to lower resolution (which is linearly dependent on the wavelength) and the ultra-short pulses were frequently not well managed through the microscope leading to lower signal levels. Users looking at samples that were perhaps only several tens of microns thick were therefore disappointed; the real advantage of multiphoton imaging is for samples greater than around 50 microns, and for *in vivo* imaging. Thus some users were purchasing the wrong instrument for their imaging challenge in the desire to be "state of the art".

The following sections do not provide rules to follow in selecting a particular methodology, but the text is written to raise questions in the minds of the potential user. Based upon the answers to these questions, the users will be able to examine in detail the capability of each method, along with the drawbacks, before making a final decision. The text uses references back to previous sections of the book where the features of a particular method have been highlighted.

14.2 BASIC SELECTION GUIDELINES

If there is one overriding rule here it is "select the most basic system that is suitable for the task in hand", in short "do not over complicate". As stated above, the latest webcam-based microscopes actually record excellent quality images. Although they may lack the flexibility of a large commercial microscope they are at a price that multiple systems can be purchased and dedicated to specific imaging tasks. In the author's own laboratory one of these systems has been converted for use as a fluorescence microscope by replacing the white LEDs with blue, and integrating a filter in front of the objective lens. This provides a simple instrument to check if a sample is well labelled before placing it onto more complex confocal, SPIM and multiphoton instruments.

Table 14.1 summarizes some of the features of each microscope described previously and thus begins to provide a starting point for instrument selection. The initial questions to be considered are:

1. What will be the source of the contrast in the image?

2. Will the sample be thin (few microns thick) such as single cells?

3. Will the sample be transparent?

4. Will optical sections be required from a thicker sample?

5. Will dynamic events be required to be observed, and at what speed?

6. What spatial resolution is actually required?

7. How much can one perturb the sample?

8. Under what conditions will the sample need to be observed (very important for biological samples that need to be kept alive and for *in vivo* imaging)?

9. For how long will the sample need to be imaged?

10. How many samples will need to be imaged?

TABLE 14.1   Features and Benefits of Various Forms of Optical Microscopy

Method	Sample Features	Advantages	Disadvantages	Details
Widefield Transmission	• Translucent or transparent sample • No labelling required • Scattering, absorption or refractive index changes for contrast • Can be dynamic	• Large field of view • Ease of use • Readily available • Low cost • High speed • Minimal perturbation	• Limited contrast mechanisms • No specific features labelled • Defocus blur • 2D images • Loss of feature in 3rd dimension	Ch 3-
Widefield reflection	• Non-transmissive samples • Surface imaging • Can be dynamic • Contrast from scattering and reflection	• Large field of view • Ease of use • Low cost • High speed • Minimal perturbation	• Limited contrast mechanisms • No specific features labelled • Defocus blur • 2D images • Loss of feature in 3rd dimension	Ch 4-
Widefield Polarization	• Transmission or reflection • Generally thin samples • Can highlight stress • Can be dynamic • Highlights birefringence	• Large field of view • Ease of use • Low cost • High speed • Minimal perturbation	• Contrast from entire thickness of sample • Limited contrast mechanisms • No specific features labelled • Defocus blur • 2D images	Ch 4-
Widefield phase	• Transmission • Generally thin samples • Can be dynamic	• Large field of view • Ease of use • Reasonable cost • High speed • Minimal perturbation • Contrast from integrated optical path length	• Contrast from entire thickness of sample • Limited contrast mechanisms • No specific features labelled • 2D images	Ch 4-
Widefield DIC	• Transmission • Generally thin samples • Can be dynamic	• Large field of view • Ease of use • Reasonable cost • High speed • Minimal perturbation • Contrast from gradient in path length	• Contrast from gradient of entire thickness of sample • Limited contrast mechanisms • No specific features labelled • 2D images	Ch 4-

(Continued)

TABLE 14.1 (CONTINUED)   Features and Benefits of Various Forms of Optical Microscopy

Method	Sample Features	Advantages	Disadvantages	Details
Widefield darkfield	• Transmission • Can be dynamic • Partially scattering sample	• Large field of view • Ease of use • Reasonable cost • High speed • Minimal perturbation • Contrast from scattering	• Limited contrast mechanisms • No specific features labelled • 2D images	Ch 4-
Widefield Fluorescence (including FLIM)	• Transmission or epi-configuration • Labelled sample • Can be dynamic • Multiple features possible • Fluorescence lifetime can show local environment	• Large field of view • Ease of use • Readily available • Good speed • Reasonable cost • Specific feature labelling	• Only labelled features visible • Sample preparation required • Generally addition of compound to sample (or genetic manipulation) • Defocus blur • 2D	Ch 4- Ch 6-
Confocal (including FLIM)	• Epi-configuration • Generally fluorescence labelled sample • Multiple features • Samples ~50–100 µm thick • Fluorescence lifetime can show local environment	• 3D Imaging to ~100 µm • Specific feature labelling • Higher resolution than widefield	• Only labelled features visible • Sample preparation required • Higher cost • Generally addition of compound to sample (or genetic manipulation) • Potentially high light levels	Ch 5- Ch 6-
SPIM	• Fluorescently labelled sample • Can be in vivo (zebrafish, *C elegans*, plants, *Drosophila*) • Generally water mounted sample • Wide field of view	• *In vivo* samples • 3D imaging • Extended time imaging • High speed possible • Reasonable cost if open SPIM, if not high cost • Minimal photobleaching	• Sample mounting • Only labelled features visible • Sample preparation required • Generally addition of compound to sample (or genetic manipulation)	Ch 7-
Multiphoton Fluorescence (including FLIM)	• Fluorescently labelled sample • Can be *in vivo* • Fluorescence lifetime can show local environment • Samples to 1–2 mm	• *In vivo* samples • 3D imaging • Extended time imaging • Significant depth possible >~1 mm	• Only fluorescent features visible • High cost • Sample preparation • Generally addition of compound to sample (or genetic manipulation) • Potentially high light levels	Ch 8- Ch 6-

(Continued)

TABLE 14.1 (CONTINUED) Features and Benefits of Various Forms of Optical Microscopy

Method	Sample Features	Advantages	Disadvantages	Details
Harmonic imaging	• Needs correct structure (e.g. collagen)	• *In vivo* samples • 3D imaging • Extended time imaging • No labelling required	• High cost • Only specific features visible	Ch 9-
Raman	• Chemically specific • Can be inherent or additional compounds • Potential for minimal perturbation to sample	• 3D imaging possible • No additional chemical required	• High cost • *In vivo* samples • 3D Imaging • Extended time imaging	Ch 10-
Holographic	• Reflection or transmission • Main contrast due to refractive index differences • 3D data in a single shot • Dynamic imaging	• Reasonable cost • High speed • Minimal perturbation to sample	• Computational power required • Complexity of image interpretation	Ch 11-
Super-resolution SIM	• Fluorescently labelled samples • Epi-configuration	• Limited 3D • Resolution to ~100 nm • Reasonable cost	• Not fast • Improved resolution but not molecular level	
Super Resolution localization (PALM, STORM)	• Fluorescently labelled samples • Epi-configuration	• Reasonable cost • Resolution to <10 nm • Potential for reasonable speed • Generally 2D	• Computational power required • Special labelling of sample	Ch 12-
Super Resolution STED	• Fluorescently labelled samples • Epi-configuration	• Limited 3D • Resolution to ~5 nm	• Very high cost • Complex to operate • Slow • Very high light levels	Ch 12-

Some of these questions are clearly interrelated and Table 14.2 begins to provide some answers on how these might guide one to the best instrument. The suggested instruments are for indication only as clearly if one has a sample which is only a few microns thick one may still wish to use a confocal to remove, for example, features that might be on the cell membrane when one wishes to look at activity in the cell nucleus.

As a general rule if one wishes to observe dynamic events the imaging method is likely to include a camera for detection. Even then one still requires sufficient signal to obtain a worthwhile image. This, for example, can be a limitation in the speed of observing some calcium signaling events. One either has to use a slower scanning multiphoton microscope to go deep into the tissue, or with a camera-based system the light levels, in thick samples, can perturb the sample either due to photobleaching leading to toxicity, or to direct phototoxicity of the sample.

14.3 SPECIALIZED TECHNIQUES

In particular in the life sciences a number of specialized techniques have been developed which provide some specific drivers towards an imaging modality. Some of these will now be discussed in outline.

14.3.1 Fluorescence Recovery after Photobleaching (FRAP)

Fluorescence recovery after photobleaching (FRAP) is a method in which an area within the sample is deliberately photobleached and then the recovery of the fluorescence monitored to provide information on cell dynamics and local diffusion rates. The typical protocol here is to image an area using a low excitation power and then to zoom in to a small area within the original field of view. This area is then imaged with a very high level of excitation for a period of time until the fluorescence intensity has been significantly reduced (different protocols use different values). The microscope is then "un-zoomed" and the original area imaged at low excitation intensity again. Over time the fluorescence returns to the area due to either the production of new fluorophores, in the case of fluorescent proteins, or the diffusion of new fluorescent molecules into the bleached area. Various mathematical models can then be applied to determine the rate of return of fluorescence.

This generally requires a beam scanned microscope (though it can be achieved using masks in the illumination optics of a widefield system) and a system with sensitive detection and the ability to take optical sections. This leads the user towards a confocal system using a relatively high power laser (to ensure rapid photobleaching as one does not want too many new molecules coming into the bleached area during the bleaching). The system should, however, also have sensitive detectors so that the imaging can be undertaken with minimal risk of bleaching while imaging. The alternative is to use a multiphoton microscope. This has the advantage that it will only bleach a single plane (the confocal illuminates and hence can bleach a cone either side of the focal plane) though the process may be slower. In either case the returning fluorescence can be monitored using a widefield camera if required, for example in high-speed diffusion situations. Full details can be found for specific protocols but the principles behind the instrument selection should be clear.

TABLE 14.2   Guideline for Microscope Selection

Question	Answer	Possible Instrument
Source of contrast	Fluorescence	Widefield fluorescent
		TIRF
		Confocal, multiphoton
		superresolution
	Refractive index (optical path)	Widefield conventional
		Phase contrast
		DIC
		Darkfield (scattering)
		Holographic (adds 3D)
	Birefringence / physical stress	Polarization
		DIC
		Holographic (potential 3D)
	Structure (overall, crystal or polymer)	DIC
		Polarization
		Harmonic
		Raman (if Raman active)
Sample thickness	Few microns	Widefield (in all forms)
		Confocal (if section really required)
		Holographic
		TIRF for surface
		Super-resolution
		Raman (chemical specificity)
	10–50 μm	Confocal (fluorescence or Raman)
		Holographic
		Multiphoton (*in vivo*)
		Harmonic
		SPIM (extended periods)
	50–1000 μm	Multiphoton
		SPIM
		Harmonic
		CARS or SRS
Transparency	Highly transparent	Widefield (in all forms)
		Confocal (depending on sample thickness)
	Scattering	Widefield (for surface only)
		Multiphoton
		Harmonic
		CARS
		Holographic
	Opaque	Reflection
Optical sectioning	No	Widefield (in all forms)
	Yes	Confocal
		SPIM
		Multiphoton
		Harmonic
		CARS
		Holographic
		Super-resolution

(Continued)

TABLE 14.2 (CONTINUED)   Guideline for Microscope Selection

Question	Answer	Possible Instrument
Dynamic events	>50 frames per second	Widefield (in all forms)
		Resonant confocal (just)
		SPIM
		Holographic (just)
	20–50 frames per second	Widefield (in all forms)
		Confocal
		SPIM
		Multiphoton (just)
		Harmonic (just)
		CARS (just)
		Holographic
	1–20 frames per second	Widefield (in all forms)
		Confocal
		SPIM
		Multiphoton
		Harmonic
		CARS
		Raman (just)
		FLIM (just)
		Holographic
	<1 frame per second	All forms above
		+ super-resolution
Spatial resolution	Features 0.2 μm or more	All forms (needs oil objective)
	Features to 0.1 μm	SIM
	5 nm to 0.1 μm	STORM
		PALM
		STED
Level of perturbation permitted	Very robust	All forms
	Some chemical tolerance	All forms including fluorescence
	Some light sensitivity	All forms except super-resolution and perhaps confocal and FLIM
	Light sensitive (in visible)	Widefield (perhaps not fluorescence)
		SPIM (possibly)
		Multiphoton
		Harmonic
		CARS
	Chemically sensitive	Widefield conventional
		Phase contrast
		DIC
		Darkfield (scattering)
		Holographic (adds 3D)
		Polarization
Sample conditions	Ambient	All forms
	Freezing	Generally all forms with special stage
	37°C	Generally all forms except perhaps super-resolution

(Continued)

TABLE 14.2 (CONTINUED)   Guideline for Microscope Selection

Question	Answer	Possible Instrument
	In vivo (whole organism)	Multiphoton SPIM Harmonic CARS
Length of imaging time	<4 hours	All forms
	4–12 hours	All forms except super-resolution
	12–24 hours	Widefield (in all forms) SPIM Confocal (with care) Multiphoton (with care) Harmonic (with care) CARS and Raman (with care)
Number of samples	Low numbers	All forms except perhaps some super-resolution methods
	High number	SPIM (with microfluidic system) Widefield (in all forms)

14.3.2 Förster Resonant Energy Transfer (FRET)

The process of Förster resonant energy transfer (FRET) was outlined in Chapter 6 Section 4.2. Briefly this is where one molecule (the donor) is fluorescently excited but is so close to another molecule (the acceptor), with an absorption at the emission wavelength of the first molecule, that energy is transferred from one to the other, very rapidly and without the actual emission of a photon. This second molecule then emits fluorescence at a longer wavelength. To monitor any change in this FRET process (which is strongly dependent on the separation of the molecules) one either needs to accurately quantify the fluorescence intensity at the two wavelengths (donor and acceptor emission wavelengths) or look at the fluorescent lifetime changes of the donor molecule.

As was discussed at some length in Chapter 13 the use of direct fluorescence quantification, which would appear to be the simplest option, is not easy. Even here, where one can measure the ratio of the two fluorescent emissions, one is looking for a change of only a few per cent in intensity levels and therefore small variations in the excitation and detection can cause significant systematic errors. As discussed in Chapter 6, measuring the fluorescence lifetime change is a more accurate method to adopt. Generally, one wishes to undertake this imaging as rapidly as possible, and one does not need an exact lifetime, therefore one only really needs to image a change in lifetime. This means that phase-based methods using a camera or beam scanned system are perfectly adequate for FRET imaging. However, the more complex time correlated single photon counting (TCSPC) method is normally employed, as this is the system that is more readily available in laboratories. This is probably one of the few examples where the simplest method of imaging is not used, with the main driver being ease of access to FLIM systems.

14.3.3 Opto-genetics, Observation of Cell Ablation and Photo-uncaging

Although these three techniques are not directly imaging methods, all need to deliver light very accurately to a specific point within a sample. An option is via an embedded optical fibre or micro-LED (more common in opto-genetic experiments) but generally the samples are placed on an optical microscope and imaged while the light activated process takes place. Generally the light to be delivered is in the blue or ultraviolet portion of the spectrum. This is most easily achieved using either a beam scanned system (typically a confocal) or with the light is introduced through the objective lens in an epi-configuration as a single spot.

The protocols are generally similar. Initially a full field image is taken and then the activation, or uncaging, light switched on. This will be a bright flash, for uncaging, or a longer lower power beam for activating opto-genetics. The resulting changes within the sample are then imaged using either a confocal, or more frequently a camera-based system. The reason for the camera is that the effects which are subsequently observed can be very rapid (for example calcium changes in the case of many uncaging experiments). Multiphoton uncaging and activation has also been used but the targeting can be more complex as one needs to be spatially correct in all three dimensions. In the case of uncaging, the cross-section for two photon activation is generally low and thus the method can be inefficient. The important aspect in the instrumentation is the ability to image, at the correct speed, and also to deliver a pulse of light to a specific targeted area.

A similar requirement is also needed in the case of cell ablation. The standard method here is to use a UV, or sometimes near infrared laser, to deliver a high burst of light to destroy the cell. This is normally delivered in an epi-configuration. However, as the move in the life sciences continues towards the use of zebrafish these experiments need to be undertaken *in vivo*. In the case of ablation within the heart this presents a challenge as the heart is beating at about three times a second. One can either watch the beating heart and manually fire the laser at what seems like the best time or, more recently, automate the process. In a method that combines real time imaging processing of a conventional widefield image, along with a predictive algorithm, cells have been ablated within a beating heart (Matrone et al. 2013). This is perhaps an example of a multidisciplinary approach in optical microscopy combining advanced real time processing to solve a technical issue in the life sciences.

An alternative method of cell ablation is through the use of a genetically encoded light activate compound KillerRed (Bulina et al. 2006). This protein requires light at around 560 nm to produce reactive oxygenation species which kills the cell. In the original work this was undertaken in cultured cells on an epi-fluorescence microscope. However, when the protein was encoded into zebrafish light activation using an epi-fluorescent microscope led to a high mortality rate within the fish. However, through the use of a modified SPIM microscope, light was introduced to the sample through the imaging objective (Buckley et al. 2017). This led to cell ablation but no fish mortality. This again illustrates that understanding the imaging, and in this case light delivery challenges, led to the selection and adaption of the best possible imaging system.

14.3.4 Imaging of Plants and Plant Cells

Plants present two significant challenges to the optical microscope. The first is that the majority of plant cells contain chloroplasts and other light sensitive and absorbing chemicals. This causes a problem both in terms of the light delivery to the sample, and also for the collection of the light. In transmission imaging this can lead to black areas within the sample where the light has been effectively absorbed by the natural compounds within the cell. This effect can frequently be overcome using near infrared transmission imaging where the absorption is significantly lower. At around 850 nm most cameras are still sensitive and high-power LEDs work well as a light source. An alternative is to use fluorescence-based techniques but utilizing fluorophores that require excitation away from the main plant chromophores.

The other complication is the presence of a cell wall, which leads to significant light scattering, again creating a problem for the light delivery and collection. The high refractive index changes due to the cell walls limit techniques such as phase contrast, DIC and polarization, though the last of these can also give some excellent images under ideal conditions. The main way around the cell wall problem is to use light with as long a wavelength as possible to minimize the scattering effects, thus non-linear microscopy has some advantages which are perhaps not currently exploited as widely as they might be.

Two routes, which perhaps have not been used as widely as they might have been are FLIM and TIRF. FLIM provides a reliable method of separating out the inherent fluorescence from the labels that may have been added, though can slow the imaging process slightly. TIRF, although only providing a narrow imaging depth, is a useful technique in this case as the light can be guided by the cell walls, enabling TIRF to operate at greater depths within plants (Johnson and Vert 2017).

SPIM has recently been shown to provide excellent extended imaging of plants during the early stages of root development (Maizel et al. 2011). Here the microscope imaging instrumentation was specially integrated into the plant propagation system with a strong emphasis on minimizing the effect that the imaging light was having on the plant growth. As SPIM provides rapid three-dimensional images over extended periods of time this method is one that can be expected to increase in its use.

Another method, which also has great potential, is CARS (Slepkov et al. 2010; Le Thuc, Yue and Cheng 2010). Here lipids within plants provide an excellent contrast mechanism and the plant sample can even be "labelled" using deuterated compounds. This is perhaps one area in which optical instrument development has lagged behind, with physicists concentrating on animal- rather than plant-based imaging methods. However, the above guidelines should help in suggesting some methods to consider within the plant sciences.

14.4 SUMMARY

It is very difficult to suggest the best way of undertaking microscopy on a particular sample but hopefully the comments above provide some guidelines to follow. As stated at the outset, the rule is to determine what you really wish to record and to use the simplest method possible to achieve those results and perturb the sample as little as possible. In particular, appreciate that light is a perturbing feature, both directly and indirectly, through chemical changes in fluorophores that have been added to the sample. Do not overlook older methods such as DIC and phase contrast for certain imaging tasks. Although fluorescence-based imaging is clearly an important tool it is not the only one available.

REFERENCES

Buckley, C., M. T. Carvalho, L. K. Young, S. A. Rider, C. McFadden, C. Berlage, R. F. Verdon, et al. 2017. "Precise Spatio-Temporal Control of Rapid Optogenetic Cell Ablation with Mem-KillerRed in Zebrafish". *Scientific Reports* 7(1): 5096.

Bulina, M. E., K. A. Lukyanov, O. V. Britanova, D. Onichtchouk, S. Lukyanov and D. M. Chudakov. 2006. "Chromophore-Assisted Light Inactivation (CALI) Using the Phototoxic Fluorescent Protein KillerRed". *Nature Protocols* 1(2): 947–53.

Johnson, A. and G. Vert. 2017. "Single Event Resolution of Plant Plasma Membrane Protein Endocytosis by TIRF Microscopy". *Frontiers in Plant Science* 8: 612.

Le, T. T., S. Yue and J-X. Cheng. 2010. "Shedding New Light on Lipid Biology with Coherent Anti-Stokes Raman Scattering Microscopy". *Journal of Lipid Research* 51(11): 3091–3102.

Maizel, A., D. von Wangenheim, F. Federici, J. Haseloff and E. H. Stelzer. 2011. "High-Resolution Live Imaging of Plant Growth in Near Physiological Bright Conditions Using Light Sheet Fluorescence Microscopy". *The Plant Journal* 68(2): 377–85.

Matrone, G., J. M. Taylor, K. S. Wilson, J. Baily, G. D. Love, J. M. Girkin, J. J. Mullins, et al. 2013. "Laser-Targeted Ablation of the Zebrafish Embryonic Ventricle: A Novel Model of Cardiac Injury and Repair". *International Journal of Cardiology* 168(4): 3913–19.

Slepkov, A. D., A. Risdale, A. F. Pegorano, D. J. Moffatt and A. Stolow. 2010. "Multimodal CARS Microscopy of Structured Carbohydrate Biopolymers". *Biomedical Optics Express* 1(5): 1347–57.

Glossary

Abbe Limit: Smallest distance that can be resolved by a conventional light microscope due to diffraction of light discovered by Ernst Abbe

Achromatic Optics: Optical elements which have the same optical power for all visible wavelengths

Airy disc: The central pattern of light produced due to diffraction from a circular aperture

Astigmatism: Optical component with greater focusing power in one direction than another

Beer-Lambert Law: Is the linear relationship between absorbance and the concentration of absorbing species

Birefringence: The optical property of a material which has a refractive index which depends on the polarization of the light

Chromatic aberration: Different wavelengths of light focusing at different points due to the variation in refractive index with wavelength

Circularly polarized light: two perpendicularly electromagnetic plane waves of equal amplitude with a 90° phase difference

Confocal Microscopy: A microscopy method in which the light from the focus of the objective lens passes through an aperture (pin hole) in the optical system such that light from outside the focus is rejected enabling "optical slices" of the sample to be recorded

Conjugate planes: Two planes in an optical system in which one point in plane "a" is re-imaged onto a point in plane "b"

Critical angle: The angle at which light starts to be totally internally reflected as it attempts to move from a high to low refractive index material

Deconvolution: A method of improving image quality by removing out of focus light computational. The software estimates (blind deconvolution) or knows the optical properties of the imaging lens and thus applies this knowledge to the images to remove "blur"

Dichromatic filter: An optical element which reflects one wavelength range (typically the shorter wavelength) and transmits the other wavelengths

Diffraction: The bending of light as it passes around the edge of an optic. A property of all waves

Dispersion: Spreading of the different wavelengths of light in space or time

Dispersion compensation: Method in which optical elements (typically prisms or gratings) are used to remove the temporal lengthening of ultra-short optical pulses caused by variations in the refractive index of materials with wavelength

Epi- configuration: The light is delivered to the sample and the image collected through the same objective lens

Field Curvature: An optical aberration that causes points on a flat object not to be focused to a common optical plane meaning some points appear out of focus within the field of view of a microscope

Fluorescence: Light is absorbed by a molecule and then re-emitted a short time later at a long wavelength

Fluorescence lifetime: The average length of time for a photon to be absorbed and a subsequent photon re-emitted by a fluorescent material

Fluorescence lifetime imaging (FLIM): The use of fluorescence lifetime as a contrast mechanism in a optical image

Förster Resonance Energy Transfer (FRET): A mechanism describing the energy transfer between two light absorbing molecules through non-radiative processes. The efficiency of the process is inversely proportional to the sixth power of the distance between the molecules

Fresnel reflection: The reflection of light at the surface where there is a change in refractive index. Around 4% for an air to glass interface for light at normal incidence

Green fluorescent protein (Gfp): An inherently fluorescent protein from the jellyfish Aequorea Victoria whose expression can be linked with another specific protein expression to fluorescently label samples. There are now multiple genetically expressed fluorescent proteins covering the complete visible spectrum

Group Velocity: The velocity of the entire pulse envelope of a short pulse of light

Infinity optics: Microscope optics in which the light from the objective is collimated as it travels through the optical system in the microscope

Interference: When the peak of one wave adds to the trough of another wave with the net result of zero (destructive interference), or when two peaks combine to give constructive interference

Köhler Illumination: The optimal illumination system for a widefield microscope producing the brightest, even illumination at the sample to provide the highest possible contrast in the sample

Linearly polarized light: All waves are oscillating in the same direction

Look-up table (LUT): The table used to display the image on a computer in which intensity is mapped to a colour, or brightness of a colour

Numerical aperture (NA): A dimensionless number that describes the angle over which an optical system can capture light. Mathematically $NA = nSin\theta$ where θ is half the cone angle of the light, and n the refractive index

Nyquist limit: This comes from a complex mathematical theory related to sampling information. In basic terms in microscopy it is the minimum spacing or resolution that is required to obtain all the information from a data set. One should sample at twice the optical resolution of a system (e.g. for a confocal scan if the axial resolution is 1 µm then the axial sections should be spaced at 0.5 µm). The same guidelines apply to temporal resolution for high-speed imaging

Optical Parametric Oscillator (OPO): Method of converting a laser light at one frequency into two output waves of lower frequency by means of second-order nonlinear optical interaction. Used in some laser systems for applications in non-linear microscopy to provide broad spectral tuning

Phase Velocity: The speed at which a wave's individual crest or trough travels through a material

PhotoActivated Localization Microscopy (PALM): A wide field super resolution methods based upon finding the centre of mass of single molecule fluorescence detected on a camera (see also STORM)

Photon: A single quanta, or particle, of light

Planck's Equation: Links the energy of a photon to its frequency or wavelength

Point Spread Function (PSF): Describes the response of an imaging system to a point source of light

Polarization: A property of a wave's direction of oscillation

Raman Scattering: The inelastic scattering of light by molecules. Generally energy is transferred to the molecule meaning the light comes out with a longer wavelength with the change in wavelength known as the Stokes shift

Randomly polarized light: A combination of waves oscillating in all directions, also known as un-polarized light

Razor Edge Filter: A dichromatic filter with a very sharp wavelength transition from reflecting to transmitting light (typically around 1 nm from 10 to 90% transmission). Used in Raman spectroscopy

Refraction (TIR): The bending of light as it enters a material of a different refractive index

Refractive index (n): Ratio of the speed of light in a material to the speed of light in a vacuum, denoted by n

Scattering: The deviation of light from a straight path due to localized variations in the refractive index of the material through which the light is propagating

Selective plane illumination microscopy (SPIM): Also known as single plane illumination microscopy or light sheet microscopy. A single thin sheet of light is used to illuminate the sample and the image is recorded orthogonally to the illumination microscopy to record a single optical slice. A 3D image can then be constructed from multiple slices

Snell's Law: Determines the level of refraction of light as it enters and leaves a material of a different refractive index

Spectral brightness: The intensity of a light source in a specific wavelength range

Spherical aberration: Loss of a sharp focus in an optical system caused by the spherical surfaces present on optical elements

Stimulated Emission and Depletion Microscopy (STED): A super resolution microscopy method where a diffraction limited area in a fluorescent sample is excited and a doughnut shaped beam used to remove the fluorescence except right at the centre of the original excitation

Stimulated Raman Scattering (SRS): Raman scattering within a sample in which the beat frequency between two laser sources is at the same frequency as a molecular vibration. This causes an change in intensity of the two laser beams

Stochastic Optical Reconstruction Microscopy (STORM): A wide field super resolution methods based upon finding the centre of mass of single molecule fluorescence detected on a camera (see also PALM)

Super Resolution Microscopy (Super-Res): Microscopy providing optical resolution beyond the Abbe diffraction limit

Time Correlated Single Photon Counting (TCSPC): Method in which the time of emission of an individual photon from a fluorescent molecule is measured compared to the original excitation light source. Provides an accurate measurement of the molecules fluorescence lifetime

Total internal reflection (TIR): Light it totally reflected at an interface as it attempts to travel from a high to low refractive index material

Total internal reflection microscopy (TIRF): A fluorescence based method in which light is guided by total internal reflection along the microscope slide and the evanescent wave excites fluorophores within around 100 nm of the glass surface. This provides super-resolution microscopy in the axial direction

Wave frequency: The number of waves passing a point in a second

Wavelength: The distance from one peak (or trough) on the wave to the next one

Widefield Imaging: A conventional microscope in which the entire field of view is imaged/recorded simultaneously

Zernike Polynomials: Mathematical series used to express optical aberration in an optical system. Each term in the series can be assigned to a specific distortion in the final image

Index

A

Abbe limit, 20
Absorption, 25
Adaptive optics, 128, 159, 217
Airy disc, 20
Astigmatism, 24
Avalanche photodiode, 36

B

Beam scanning acousto-optic, 77
Beam scanning galvanometer, 75
Beam scanning resonant, 77
Beer-Lambert law, 26
Birefringence, 56

C

CARS, *see* Coherent Anti-Stokes Raman Scattering
CARS microscopy, 185, 249–251, 253
Chromatic aberration, 22
Cleaning optics, 49
Coherent Anti-Stokes Raman Scattering
    (CARS), 179
Confocal microscopy, 73, 246, 249–251
Conjugate planes, 42
Critical angle, 19

D

Dark field microscopy, 26, 65, 246, 249, 250
Deconvolution, 240
DIC Microscopy, 27, 62, 72, 245, 249, 250
Diffraction, 19
Digital cameras, 32, 50, 52, 199
Dispersion, 16
Dispersion compensation, 146

F

Field curvature, 24
Fluorescence, 27, 97, 176

Fluorescence filter selection, 70, 79
Fluorescence lifetime, 28, 97, 98
Fluorescence microscopy, 68, 72, 97, 118, 139, 217,
    246, 248–250, 252
Förster Resonance Energy Transfer (FRET), 97,
    114, 251
Fresnel reflection, 18
FRET, *see* Förster Resonance Energy Transfer

G

Group velocity, 16

H

Harmonic microscopy, 165, 247, 249
Holographic microscopy, 193, 247, 249

I

Illumination
    arc lamp, 29
    filament, 29
    laser, 30, 142
    LED, 29
    Mercury, 29
    metal halide, 29
Image processing, 225
Image speckle, 231
Infinity optics, 41
Interference, 19
Inverted microscope, 40, 86

K

Köhler illumination, 43

L

Localization microscopy (STORM and PALM), 217

M

Multiphoton microscopy, 130, 135, 139, 243, 246, 249, 250

N

Nyquist limit, 51, 84, 129, 133, 226

O

Objective lens, 46, 87, 152, 189
Optical parametric oscillator, 185
Opto-genetics, 252

P

Phase contrast microscopy, 27, 58, 72, 245, 249, 250
Phase velocity, 16
Photodiode, 36, 161
Photomultiplier, 34, 81, 103
Pixel binning, 33, 52, 58, 188
Planck's law, 15
Polarization, 16
Polarization circular, 16
Polarization linear, 16
Polarization microscopy, 55, 72, 245, 249, 250
Polarization random, 16

R

Raman microscopy, 112, 139, 175, 181, 247, 249–251
Raman scattering, 28, 175, 176
Rayleigh criterion, 21

Refraction, 17
Refractive index, 15, 17

S

Scattering, 25
Selective Plane Illumination Microscopy (SPIM), 117, 246, 249, 250
Shot Noise, 231
Single Photon Avalanche Photdiodes (SPAD), 36, 103
Snell's Law, 18, 210
SPAD, *see* Single Photon Avalanche Photdiodes
Spherical aberration, 22
SPIM, *see* Selective Plane Illumination Microscopy
Spinning disc confocal, 82
Stimulated Emission and Depletion Microscopy, 220, 247, 250
Structured Illumination Microscopy, 212, 247, 250
Super resolution microscopy, 207, 247, 249, 250

T

Time Correlated Single Photon Counting, 101
Total internal reflection, 18, 203
Total Internal Reflection microscopy, 203, 208, 249
Two photon microscopy, 28, 139, 244, 249

U

Upright microscope, 40

Z

Zernike polynomials, 21